GRADE 8
MATHEMATICS

New York Reviewers
Harley • Shatz

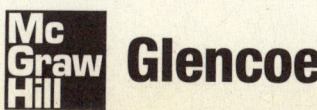

McGraw Hill **Glencoe**

New York, New York   Columbus, Ohio   Chicago, Illinois   Peoria, Illinois   Woodland Hills, California

New York Reviewers

Debra L. Harley
Mathematics Coordinator K-12
Lindenhurst Public Schools
Lindenhurst, New York

Erica J. Shatz
Former Director of
  Mathematics CSDI NYC
Adjunct Professor,
  Brooklyn College and
  Hunter College
New York, New York

Photo Credits: Title Page (l) Dennis Degnan/Index Stock Imagery; (cl) CORBIS; (cr) Alan Schein Photography/CORBIS; (r) CORBIS

The McGraw·Hill Companies

Send all inquiries to:
The McGraw-Hill Companies
8787 Orion Place
Columbus, OH 43240-4027

ISBN: 0-07-874394-X

New York Review Series, Grade 8 Mathematics

5 6 7 8 9 10 009 13 12 11 10 09 08

Contents

Unit 1: Review of Grade 7 Performance Indicators (Post-March)

Chapter 1: Grade 7 Algebra

Chapter 2: Grade 7 Geometry

Chapter 3: Grade 7 Measurement

Unit 2: Grade 8 Performance Indicators (Pre-March)

Chapter 4: Grade 8 Number Sense and Operations

Chapter 6: Grade 8 Geometry

Chapter 7: Grade 8 Measurement

How to Master the New York Test

What is the New York Test?

In March, you will be taking the New York State Test in Mathematics for Grade 8. The test covers the topics and skills that you will be learning at school this year from September to March. It also covers the topics you were taught last spring in your seventh-grade math class.

Your performance on the test will show how well you have learned the material that is required of all eighth-grade students in the state. It can also help you and your teacher identify which areas and skills you are strong in, and which you may need to focus on and develop.

There are three types of questions that you will see on the test:
- Multiple Choice
- Short Response
- Extended Response

For multiple-choice questions, you will need to choose the correct answer to a problem from four given choices. Short- and extended-response questions, however, ask you to solve a problem and write your own answer. You may also need to show your work by writing out each step in your calculations, drawing a picture, or describing in words how you solved the problem.

The test is divided into three sessions that are stretched out over two days. Here is a breakdown of the type and number of questions that are included in each test session.

	Session	Multiple Choice	Short Response	Extended Response	Total Questions	Time
Day 1	1	27	–	–	27	45 min
	2	–	4	2	6	35 min
Day 2	3	–	8	4	12	65 min
	Total	27	12	6	45	145 min

Knowing what kinds of questions to expect on a test can help you to prepare and do your best. The following examples demonstrate the three types of questions you will see on the New York State Test, and some tips that you can use to solve each.

Multiple-Choice Questions

All of the questions in Session 1 are multiple choice. For these questions, you will be asked to choose the correct answer to a problem from four given choices. Calculators are NOT allowed during this part of the test.

Follow these steps to help you answer this type of question:

1. **Read the question carefully and look over any graphic that may be included.**
2. **Read all the answer choices and cross out any that you know are wrong.**
3. **Solve the problem.**
4. **Check your answer.**
5. **Make sure you clearly mark the correct answer on your scorecard.**

EXAMPLE

1 A tutor charged $62.50 for 2.5 hours of work. What is the tutor's hourly rate?
  A $5 per hour
  B $15 per hour
  C $25 per hour
  D $50 per hour

You need to find how much the tutor charges for 1 hour of work. To solve the problem, you can divide $62.50 by 2.5.

Now look at the answer choices. You know that answer A cannot be correct, because $5 is too small of an amount. Answer D is also incorrect because $50 is too high of a number. After eliminating these choices, the only possible answers are B and C.

Use scrap paper to solve the problem: $62.50 \div 2.5 = 25$, so the correct answer is C, $25 per hour.

Check your answer by multiplying: $25 \times 2.5 = 62.5$ or $62.50, so your answer is correct. On your scorecard, fill in the circle for answer C.

Tips
  • If you do not understand the question or cannot eliminate any of the answer choices, move on to the next question. You can come back to this question when you finish the rest of the questions in the section.
  • If you can eliminate two of the answer choices but cannot solve the problem, then make your best guess. No points will be taken off your score for a wrong answer, and you have a good chance of guessing correctly!

Short-Response and Extended-Response Questions

All of the questions in Sessions 2 and 3 will be short or extended response. These questions will ask you to solve a problem and write your own answer on the line(s) provided. You may also need to show your work by writing down each step in the problem, drawing a picture, completing a chart, or explaining in words how you found the answer.

The only difference between short- and extended-response questions is that extended-response questions have two or more parts. Calculators ARE allowed during these parts of the test.

Follow these steps to help you answer these types of questions:

1. **Read the question carefully and look over any graphic that may be included.**
2. **Pick the best method for solving the problem.**
3. **Solve the problem.**
4. **Check your final answer and write it on the given line(s) or space.**
5. **Make sure that you show your work and that you have answered each part of the question.**

SHORT–RESPONSE EXAMPLE

1 The area of the shaded square below is half the area of the whole rectangle. What is the area of the rectangle?

12 cm

Show your work.

Answer _____ square centimeters

For this question, you are given a diagram to use. The diagram gives you the length and width of the shaded square, which is half the area of the whole rectangle. To solve the problem, you can find the area of the shaded square and multiply it by 2.

You can use your calculator to multiply. However, you must write out the steps in the given space to show how you found your answer. It is also a good idea to label your work, as shown.

Show your work.

area of shaded square
↓
$(12 \times 12) \times 2 = 288$

Is the area of the rectangle 288 square centimeters? The problem tells you that the length of the square is half the length of the rectangle. You can check your answer by doubling the length of the square and using the new measurement to find the area of the rectangle.

length of rectangle = 12 × 2 = 24
area of rectangle = 24 × 12 = 288

Your answer is correct, so write 288 on the line labeled *Answer*. Notice that the units are already written for you.

The same question could also be used as the first part, or *Part A*, of an extended-response question. In this case, the second part of the extended-response question would be labeled *Part B*, as below. The example below shows how you could answer this part of the question.

EXTENDED-RESPONSE EXAMPLE

Part B

On the lines below, explain how you found your answer.

The diagram shows that each side of the shaded square is 12 centimeters. Since the square is half the area of the whole rectangle, I multiplied 12 times 12 to find the area of the square, 144. Then I multiplied the area of the square by 2 to find the area of the rectangle: 144 x 2 = 288.

Tip

- Even if you do not get the correct final answer, you may be given points for the work you have done. For this reason, it is VERY important that you write or draw as neatly as possible when showing your work.

Test-Taking Tips

When you take any test, there are some things that you should always do:

1. Try to get a full night's sleep before the test, so you are well rested.
2. Read the directions carefully. If you do not understand what you are supposed to do, ask your teacher to explain before the test begins.
3. Read each question or problem carefully.
4. Manage your time for each section. If you are not sure how to solve a problem, skip it. Then go back to it once you have finished the other problems in the section.
5. Check your answer for each problem.

How to Use This Book

This book will help you review and develop the math skills you need to succeed on the New York State Test in Mathematics for Grade 8. The extra practice that you will gain from going over the solved examples and doing the problems will also help you to do better in your schoolwork throughout the year.

Unit 1 is a review of the seventh-grade topics that may be included on the test. Unit 2 includes all of the eighth-grade material that you will be expected to know. However, the lessons are written so that you can review them in any order. Using the table of contents, you can easily find the lessons that focus on the topics you need the most help with.

Every lesson has the following features:

- A list of skills or performance indicators, including the process skills that you will need to apply to the problems.
- Any vocabulary and definitions that you will need to know for the lesson.
- Basic facts and background information on the topic, to help you get up to speed.
- Two or three example problems, solved step-by-step.
- *Understanding the Solution:* Explanations of the answers for each solved example, or how to check your answer.
- *Try It:* Example problems that you can solve on your own using the same methods.
- *Exercises:* Multiple-choice, short-response, and extended-response practice questions that are like the questions you may find on the test.

Special *Problem-Solving Lessons* are also included in each chapter. These shorter lessons help you to develop and use different methods for solving problems.

Each chapter and unit in the book is also followed by a sample practice test. These tests provide extra practice in the different ways that each topic may be tested.

The next page contains a list of formulas and other information that you will need to know and use to solve problems. Use this sheet to help you identify and review important concepts.

A paper ruler and protractor are also included at the end of the book, so you can cut them out and practice using them to answer problems. These tools are similar to the ones that you will be expected to use on the test.

Mathematics Reference Sheet— Grade 8

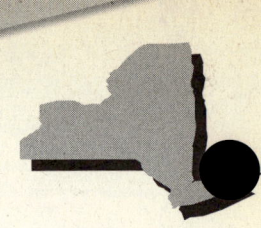

Conversions

Temperature Conversion

$F = \frac{9}{5}C + 32$

$C = \frac{5}{9}(F - 32)$

Measurement Conversion

1 mile = 5,280 feet

1 yard = 3 feet

Formulas

Pythagorean Theorem

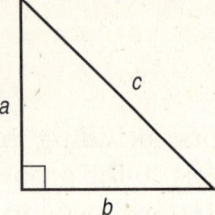

$c^2 = a^2 + b^2$

Simple Interest $I = prt$

Distance Formula $d = rt$

Slope-Intercept Form $y = mx + b$
$m = slope$ and $b = y\text{-}intercept$

Understanding Variables and Algebraic Expressions

New York Performance Indicators

7.A.2 Add and subtract monomials with exponents of one

7.A.3 Identify a polynomial as an algebraic expression containing one or more terms

7.PS.3 Understand and demonstrate how written symbols represent mathematical ideas

7.PS.6 Represent problem situations verbally, numerically, algebraically, and graphically

VOCABULARY

A **variable** is a symbol or letter that represents an unknown in an expression.

A **monomial** is a number, variable, or product of numbers and variables.

A **polynomial** is an algebraic expression that contains one or more monomials, also called **terms**.

Like terms have variables that are the same, such as $5y$ and $4y$. Numbers that are not multiplied by variables, such as 3 and 11, are like terms called **constants**.

REVIEW

Understanding Algebraic Expressions

An expression is algebraic if it contains operations with variables and numbers. Look at the algebraic expression below.

$$8a + 5a - 7 - 2b$$

This expression is a polynomial. $8a$, $5a$, 7, and $2b$ are monomials. $8a$ and $5a$ are like terms that have the same variable, a. After combining the like terms, the polynomial will have three terms: $13a - 7 - 2b$.

Applying Algebraic Expressions

Rhonda and her mother rode the subway to Times Square. Rhonda paid half the full fare with a student discount. Her mother paid the full fare. Write an algebraic expression that shows the combined cost of both fares.

Let x represent the full fare.

$$x + \frac{1}{2}x$$

Simplifying an Expression

You can often simplify polynomials by combining like terms. You can also combine any constants in the expression. The resulting expression is equivalent to the original, but it has fewer terms.

EXAMPLE 1

Sheila is putting one fence up around her vegetable garden and another fence around her whole backyard. What expression shows the total amount of fencing Sheila will need?

You can write and simplify an algebraic expression.

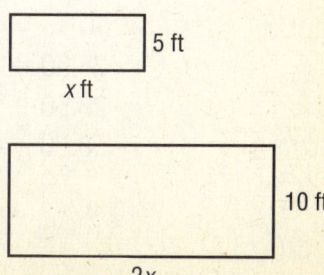

Write an expression to represent the total amount of fencing. You must add the two perimeters or total distance around each rectangle.

Total fencing $= (x + x + 5 + 5) + (2x + 2x + 10 + 10)$

Simplify the expression by combining like terms.

$$= (x + x + 2x + 2x) + (5 + 5 + 10 + 10)$$
$$= 6x + 30$$

▶ **Understanding the Solution**   The amount of fencing Sheila needs is given by the expression $6x + 30$, which is a polynomial. This simplified expression is equivalent to the original expression, $(x + x + 5 + 5) + (2x + 2x + 10 + 10)$.

TRY IT!

Richard buys 3 bike helmets for his children and 2 for himself and his wife. All of the helmets are the same price. He also buys one new bike seat for $25. Write an expression that shows how much Richard spent.

Evaluating an Expression

You can evaluate or find the value of an algebraic expression by replacing the variables with their given values.

EXAMPLE 2

Brian needs to rent a few tools for a project. A power drill rental costs a flat fee of $10 and an hourly rate of $2. A circular saw rental costs a flat fee of $15 and an hourly rate of $3. Brian rents both tools for 6 hours and buys a drill bit for $3.50. What is the total cost?

You can solve this problem by writing and evaluating an algebraic expression.
Write an expression to represent the situation. Let $h = $ number of hours.

Total cost $= (10 + 2h) + (15 + 3h) + 3.50$

Simplify the expression: combine the like terms and combine the constants.

$$= 28.50 + 5h$$

Evaluate the expression by substituting 6 for h.

$$= 28.50 + 5(6)$$
$$= 28.50 + 30$$
$$= 58.50$$

▶ **Understanding the Solution**   The expression $28.50 + 5h$ equals $58.50 only if $h = 6$. Say Brian rented both tools for 4 hours instead of 6 hours. You can evaluate the same expression for $h = 4$ and get a different total cost: $28.50 + 5(4) = 28.50 + 20 = 48.50$.

TRY IT!

Cassie is visiting the amusement park at Coney Island. She rides the Cyclone roller coaster twice, the Tilt-a-Whirl once, and the bumper cars once. Then she spends $9 on dinner and $5 on a souvenir. Let r = rides taken. Write an expression for how much Cassie spent. If the rides cost $5 each, what is the total cost of her trip?

EXAMPLE 3

Jack is shopping for school supplies. He picks up 5 folders and 3 notebooks for his classes and 2 extra folders. He also picks out 17 book covers for $1 each, but puts back 12 of them. If folders cost $2 each and notebooks cost $3 each, how much did Jack spend on school supplies?

You can solve this problem by writing and evaluating an expression with two variables. Let f = the number of folders and n = the number of notebooks. Write an expression for the total cost, or amount spent.

Total cost $= 5f + 3n + 2f + 17 - 12$

Simplify the expression by combining like terms and combining constants.

$$= 7f + 3n + 5$$

Evaluate the expression by substituting 2 for f and 3 for n.

$$= 7(2) + 3(3) + 5$$
$$= 14 + 9 + 5$$
$$= 28$$

▶ **Understanding the Solution**   There are no like terms in the expression $7f + 3n + 5$, so you cannot simplify further. The variables f and n each have a different value because the cost of folders and notebooks is different.

TRY IT!

Lindsey puts 4 apples and 3 bananas in her grocery basket. She puts 2 apples back, but picks up a carton of orange juice for $0.99. If apples cost $0.20 each and bananas cost $0.15 each, how much did Lindsey spend at the grocery store?

Exercises

1 A health club has an initial membership fee of $120 and monthly dues of $49. Use the variable *x* to write an expression for the total cost of joining the health club for 9 months. Identify the variable. Then find the total cost of joining for 9 months.

Show your work.

Answer _____

MULTIPLE CHOICE

2 What is the value of $4x + 12 - y$ if $x = 5$ and $y = 10$?

A 16

B 22

C 32

D 47

3 Gavin's bedroom is *x* feet long by *y* feet wide. If $x = 12$ and $y = 10$, use the expression $2x + 2y$ or $2(x + y)$ to find the perimeter.

F 22 feet

G 32 feet

H 34 feet

J 44 feet

4 Which expression is the equivalent of $3x + 2y + 8 - 3$?

A $5xy + 5$

B $3x + 2y + 5$

C $6xy + 8 - 3$

D $5y - 5$

5 What is the value missing from the table?

x	x − 21
25	4
31	9
35	?

F 26

G 22

H 14

J 13

6 A parking lot charges a flat rate and $4 for every hour. The cost of parking the car is $4x + 5$. In the expression, what represents the flat fee?

A 4

B 5

C *x*

D $4x$

7 Which expression has exactly three like terms?

F $9a + 2a - 3a$

G $2 - 18y + 2y$

H $z + 2a + c$

J $2a - 7d - 15a$

8 Carl brought his car to be repaired and was charged $210 for parts and $40 an hour for labor.

Part A

Write an algebraic expression that shows the cost of the repairs. Indicate what the variable in the expression represents in your work.

Show your work.

Expression _____

Part B

It took the mechanic from 9 A.M. to 1 P.M. to fix Carl's car. What was the total cost of the repairs?

Show your work.

Answer _____

LESSON 1.2 Solving Equations

 New York Performance Indicators

7.A.4 Solve multi-step equations by combining like terms, using the distributive property, or moving variables to one side of the equation

7.PS.1 Use a variety of strategies to understand new mathematical content and to develop more efficient methods

7.PS.15 Choose methods for obtaining required information

VOCABULARY

An **equation** is a mathematical statement with an equals sign.

An equation can be **true**, **false**, or an **open sentence**, if it has one or more variables.

The **solution** of an equation is a value that makes the equation true.

REVIEW

Understanding Equations

An equation shows that two amounts are equal in value.

$$x = 8$$

To keep the amounts on both sides of the equals sign equal, you must perform the same operation on both sides.

Examples
Add 3 to both sides: $x + 3 = 8 + 3$
Subtract 1 from both sides:
$x - 1 = 8 - 1$
Multiply both sides by 2: $2x = 16$
Divide both sides by 4: $\frac{x}{4} = \frac{8}{4}$

Applying Equations

The annual budget for Saratoga National Historical Park in 2004 was $1,605,000. This was a decrease of about $22 per square acre from 2002, when the budget was $1,679,000. What is the approximate area of the park in square acres?

Let x = the number of square acres.

$$(1,679,000 - 1,605,000) \div x = 22$$
$$74,000 \div x = 22$$

This equation is equivalent to

$$x = \frac{74,000}{22}$$
$$x \approx 3,364 \text{ acres}$$

Checking a Solution

You can check a solution to an equation to see if it makes the statement true.

EXAMPLE 1

Is the equation $10 + 6x = x - 5$ true for $x = -3$?

You can solve the problem by substituting −3 for x.

Substitute −3 for x in the equation. If the expressions on both sides of the equals sign have the same value, then the equation is true for $x = -3$.

$$10 + 6(-3) = (-3) - 5$$
$$10 + (-18) = -8$$
$$-8 = -8$$

Since −3 made the equation $10 + 6x = x - 5$ true, −3 is a solution of the equation.

▶ **Understanding the Solution** The equation $10 + 6x = x - 5$ has only one solution. Except for $x = -3$, the equation is not true.

TRY IT!

Is 4 a solution for $30 - 15b = 2 + 2b$? Why or why not?

Solving an Equation

To solve an equation with a variable, you must move the variable terms to one side of the equals sign. To isolate or move the variable, you can often combine like terms, use inverse or opposite operations, and apply the distributive property.

EXAMPLE 2

A taxi ride costs $2.50 plus $2 for each mile. There is a surcharge of $0.50 for all rides between 8 P.M. and 6 A.M. Trinh and her mother took a taxi home from the shopping mall at 10 P.M. and paid a total of $18. How far does Trinh's family live from the shopping mall?

You can solve this problem by writing and solving an algebraic equation.

Write an equation to represent the situation. Let x represent the unknown number of miles.

$$2.50 + 2x + 0.50 = 18$$

Simplify the equation by combining the constants and like terms.

$$2x + 3 = 18$$

Subtract 3 from both sides.

$$2x + 3 - 3 = 18 - 3$$
$$2x = 15$$

Divide both sides of the equation by 2, which is the same as multiplying by $\frac{1}{2}$.

$$\left(\tfrac{1}{2}\right)2x = 15\left(\tfrac{1}{2}\right)$$
$$x = 7.5 \text{ miles}$$

▶ **Understanding the Solution** Check your answer by substituting 7.5 for x in your original equation. $2.50 + 2(7.5) + 0.50 = 2.50 + 15 + 0.50 = 18$, the cost of the taxi ride, so 7.5 miles is the correct answer. Always use the original equation to check your answer, in case you made a mistake when simplifying.

TRY IT!

What is the value of y if $4y - 5 + y + 2 = 22$?

EXAMPLE 3

Nancy roller skates from her home at a steady rate of 5 miles per hour. Two hours later, her sister Karen leaves home and bicycles along Nancy's route at a rate of 8 miles per hour. How long after Nancy leaves home will Karen catch up to her?

You can solve this problem by writing and solving an algebraic equation.

Rate $= \dfrac{\text{distance}}{\text{time}}$, so distance = rate $\times$ time.

Nancy's distance $= \dfrac{5 \text{ miles}}{\text{hour}} \times x \text{ hours} = 5x$

Karen's distance $= \dfrac{8 \text{ miles}}{\text{hour}} \times (x - 2) = 8(x - 2)$

Let x represent Nancy's time and $x - 2$ represent Karen's time.

The two sisters will meet when both have traveled the same distance.

Now set both distances equal and solve for x.

$$5x = 8(x - 2)$$

Use the distributive property to simplify.

$$5x = 8x - 16$$

Add 16 to both sides of the equation.

$$16 + 5x = 8x - 16 + 16$$
$$16 + 5x = 8x$$

Subtract $5x$ from both sides of the equation.

$$16 + 5x - 5x = 8x - 5x$$
$$16 = 3x$$

Divide both sides of the equation by 3, which is the same as multiplying by $\frac{1}{3}$.

$$\left(\tfrac{1}{3}\right)16 = \left(\tfrac{1}{3}\right)3x$$

$$\frac{16}{3} = x \text{ or } x = 5\tfrac{1}{3}$$

▶ **Understanding the Solution** Nancy's sister will catch up about 5 hours after Nancy leaves. In 5 hours, Nancy covers about 25 miles. Karen leaves 2 hours later but still can cover about 24 miles in 3 hours. So the answer is reasonable.

TRY IT!

Two cars leave Syracuse an hour apart. Car A leaves first and travels at a constant rate of 50 miles per hour. Car B leaves an hour later and travels at a constant rate of 60 miles per hour in the same direction. When will Car B pass Car A?

Exercises

SHORT RESPONSE

1 Matthew's cell phone plan charges $15 a month and 25 cents for every minute. If his bill for one month is $30.75, how many minutes did he use? Let x = the unknown number of minutes. Write and solve the equation.

Show your work.

Answer _____ minutes

MULTIPLE CHOICE

2 Which is a solution to the equation $3a + 6 = 9a + 8$?

A 6

B 4

C 2

D $-\frac{1}{3}$

3 The formula for density D is mass M divided by volume V $\left(D = \frac{M}{V}\right)$. The density of a mineral is 4 g/cm³. If Jake has 12 grams of the mineral, what is the volume?

F 3 cm³

G 4 cm³

H 16 cm³

J 48 cm³

4 Brittany saves the same amount of money each week. She had $250 eight weeks ago. After withdrawing $10, her current balance is $400. How much was she saving weekly?

A $8

B $20

C $25

D $50

5 Solve the following equation.
$4y = 3(y + 11)$

F 3

G 4

H 11

J 33

6 A rectangle has the dimensions shown.

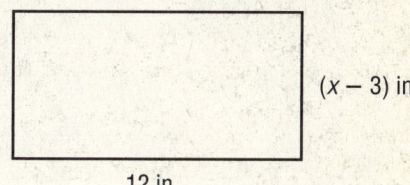

$(x - 3)$ in.

12 in.

The area of the rectangle is 24 square inches. What is the value of x?

A 5

B 4

C 3

D 2

7 Which of the following equations is ***not*** equivalent to $x + 8 = 11$?

F $(x + 8) - 3 = 8$

G $(x + 8) + 5 = 16$

H $2(x + 8) = 24$

J $x = 3$

8 Mike leaves school at 2 P.M. and bikes home at 6 miles per hour. His brother leaves school a half hour later and follows Mike's route home at 12 miles per hour. What is the best estimate of the time his brother will pass him?

Show your work.

Answer _____ P.M.

Identifying Patterns in Equations

VOCABULARY

A **sequence** is an ordered list of numbers that follow a pattern.

Each number in a sequence is called a **term**.

The **rule** for a pattern explains how to find each term in the sequence.

REVIEW

Understanding Patterns

You can represent a pattern using a sequence, a table of numbers, or drawings. The pattern shown below is to start with 1 and add 2 to each term.

Term	Number
1	1
2	3
3	5
4	7

The table corresponds to the sequence 1, 3, 5, 7, ...

Since the order of the terms is important in a sequence, each term is numbered as shown in the left column. For example, the third term in the sequence is 5.

Applying Patterns

You can use the rule for a pattern to find the next term. To find the rule, see how the figures or numbers change.

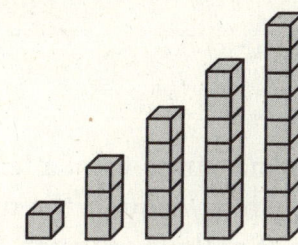

Notice that all the terms in the sequence are odd numbers. They can be represented by $2n - 1$, where n is the term number. For example, $n = 5$ for the fifth term.
Fifth term = $2(5) - 1 = 9$

Identify a Pattern

You can solve a problem by drawing a representation of the pattern to identify the pattern.

EXAMPLE 1

What is the fifth term in the table?

Term	1	2	3	4	5
Number	12	9	6	3	?

You can solve this problem by drawing a graphic representation of the pattern.

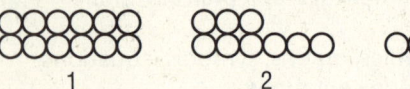

Notice that each term is three less than the previous term. The last term should have 0 circles because $3 - 3 = 0$.

▶ **Understanding the Solution** By subtracting 3 from the previous term, you can find the next term. Check that $12 - 3 = 9, 9 - 3 = 6, 6 - 3 = 3$, and $3 - 3 = 0$.

TRY IT!

Draw the fourth term of the pattern to find the missing number.

Term	1	2	3	4
Number	4	7	10	?
Drawing				

EXAMPLE 2

The amount of water that drips from a leaky faucet can be represented by the equation $3h - 1 = g$, when h equals the number of hours that have passed and g equals the number of gallons of water. Draw a graphic representation of this pattern to show how many gallons will drip from the faucet in 5 hours.

You can use the equation to help you draw the first five terms in the pattern.
In your drawing, let each water droplet represent 1 gallon of water. The first term should show how many gallons drip in 1 hour.

$$h = 1 \longrightarrow 3h - 1 = 3(1) - 1 = 2 \text{ gallons}$$

$$h = 2 \longrightarrow 3h - 1 = 3(2) - 1 = 5 \text{ gallons}$$

Two gallons of water drip in the first hour, and three additional gallons drip in the following hour.

Continue evaluating the expression $3h - 1$ for $h = 3, 4,$ and 5.

```
              O           O O        O O O       O O O O
    O         O O         O O O      O O O O     O O O O O
    O         O O         O O O      O O O O     O O O O O
    1           2            3           4            5
```

The graphical pattern shows that 14 gallons are lost by the end of the fifth hour.

▶ **Understanding the Solution**   You can check your answer in the equation $3h - 1 = g$ when $h = 5$. Substitute 5 into the equation and you can see that 14 gallons will be lost in 5 hours.

TRY IT!

A savings account begins with $15 and earns $0.25 a week. How much money is in the account after 12 weeks? Use the equation $15 + 0.25w = t$ when w equals the number of weeks and t equals the total in the account.

Exercises

SHORT RESPONSE

1 A quilter is adding blocks to a quilt pattern. The pattern started wih 5 blocks, and with each new color, she adds on 3 blocks. How many blocks will the pattern have after she has added 4 colors? Use a graphical representation to show the pattern.

Show your work.

Answer _____ blocks

2 What is the fifth term in the table?

Term	Number
1	100
2	88
3	76
4	64
5	?

A 62
B 54
C 52
D 50

3 What is the fifth term in the pattern?

F

G

H

J

4 Sarah thinks that her hair grows at the rate of half an inch per month. How long will her hair grow in 6 months?

A 0.5 inches
B 1.5 inches
C 3 inches
D 12 inches

5 Nicholas is creating four designs with tiles behind the kitchen sink. How many tiles will be in the fourth design?

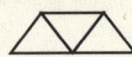 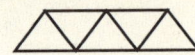

F 2
G 5
H 7
J 9

6 Which graphic representations represent the next two numbers in the pattern?

Term	1	2	3	4	5
Number	36	25	16	?	?

A

B

C

D

7 Amanda is selling jewelry on-line. The table records how many e-mails she receives each day.

Day	1	2	3	4
E-mails	1	3	9	27

Which best describes the rule for this pattern?

F multiply by 3
G add 3
H multiply by 3 and add 1
J add 18

EXTENDED RESPONSE

8 What is the fifth term in the table?

Term	1	2	3	4	5
Number	16	8	4	2	?

PART A

Use a graphic representation to show the pattern and find the fifth term.

Show your work.

Answer ____________

PART B

Use words to describe the pattern. How do you get each number in the sequence?

____

____

____

LESSON 1.4 Problem-Solving Strategy: Solving Word Problems

New York Performance Indicators

8.PS.14 Determine information required to solve the problem
8.PS.16 Justify solution methods through logical argument

How do you solve a problem?

There are four important steps that you should always follow.

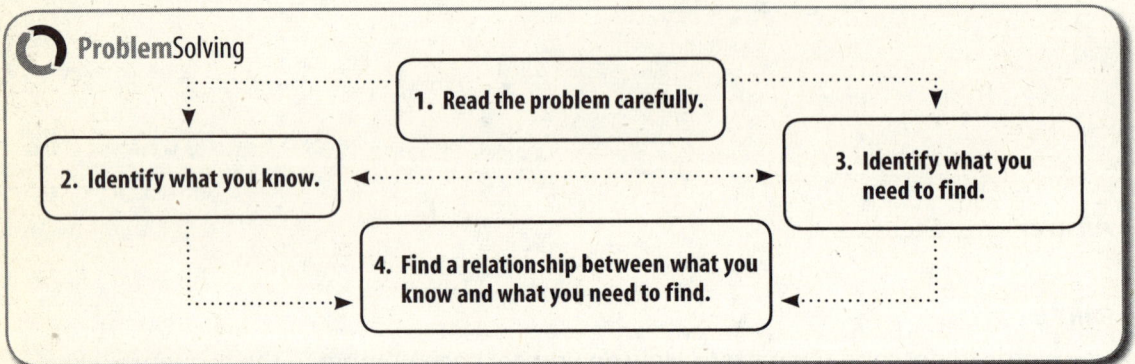

Problem: Sara worked six weeks in the summer and saved $750. Each week, she earned the same amount, spent $45 of her pay, and saved the rest. How much did Sara make each week?

SOLUTION

What do you know?

Total savings in 6 weeks: $750

She earned the same amount each week.

Weekly spending: $45

What do you need to find?

Her weekly pay.

Find the relationship.

How would you find the weekly pay? The problem gives you the clues.

$$\text{Weekly pay} = \text{Weekly spending} + \text{Weekly savings}$$

You know the weekly spending, but you also need to know the weekly savings. Did Sara save the same amount each week? The problem doesn't say that.

Sara earned the same amount each week and spent the same amount each week. From this, you can conclude that she saved the same difference.

$$\text{Weekly Savings} = \frac{\$750}{6} = \$125$$

$$\text{Weekly Pay} = \$45 + \$125 = \$170$$

▶ **Understanding the Solution** Always ask whether the solution makes sense.
Sara made $170 each week and spent only $45 each week.

$$\$170 - \$45 = \$125 \text{ and } \$125 \times 6 = \$750$$

1 Lenox is planning to read 3 books in 5 weeks. The average number of pages in each of the 3 books is 184. He wants to read the same number of pages every day except on Friday, when he doesn't plan to read. How many pages should he read per day to finish the 3 books in 5 weeks?

Show your work.

Answer _____ pages

2 Carla's driver's license was suspended because she had 23 points on her record. All of the points came from speeding tickets. She had one ticket for driving 8 miles over the speed limit and one for driving 13 miles over. Then she got two more tickets worth the same number of points each. Using the table below, how many miles per hour over the speed limit was Carla driving when she got the last two tickets?

New York Driver Violation Point System

MPH Over the Speed Limit	Points
1–10	3
11–20	4
21–30	6
31–40	8
40+	11

Show your work.

Answer _____ mph

LESSON 1.5 Creating Algebraic Patterns

New York Performance Indicators

7.A.8 Create algebraic patterns using charts/tables, graphs, equations, and expressions

7.PS.4 Observe patterns and formulate generalizations

7.RP.8 Apply inductive reasoning in making and supporting mathematical conjectures

7.CN.4 Model situations mathematically, using representations to draw conclusions and formulate new situations

---

VOCABULARY

In an **algebraic pattern**, the same operation(s) is performed on each number to get the next number in the sequence.

REVIEW

Understanding Patterns

The pattern for the sequence below is to add 3 to each number.

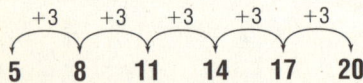

The next term in the sequence is 20 + 3, or 23.

Creating Algebraic Patterns

You can use algebra to write a rule for a pattern.

5, 8, 11, 14, 17, 20, ___

The value of the nth term is

$5 + (n - 1)3$.

When $n = 7$, the 7th term is

$5 + (7 - 1)3 = 23$.

Finding and Using a Pattern

You can find the rule for a pattern and use it to find a missing value, predict the next number in the sequence, or fill a table of values.

EXAMPLE 1

Tyshan is selling tickets for his school play. People who buy more than one ticket get a discount. The table to the right shows the cost of different numbers of tickets. How much will it cost to buy 11 tickets?

First look for a pattern.

Tickets	Cost ($)
1	19
2	26
3	33
4	40
5	47

Begin by looking at how the ticket prices change. Subtract to find the difference between 26 and 19: 26 − 19 = 7. Will adding 7 to 26 give you 33? Will it give you each number in the table?

Tickets	Cost ($)	Add Multiples of 7
1	19	
2	26	$19 + 1(7) = 26$
3	33	$19 + 2(7) = 33$
4	40	$19 + 3(7) = 40$
5	47	$19 + 4(7) = 47$

The pattern is to add multiples of 7 to the first term.

Now use the pattern rule to find the answer.

To find out how much 11 tickets would cost, you can continue the pattern by adding 10 times 7 to the first term.

$$19 + 10(7) = 89$$

It will cost $89 for 11 tickets.

▶ **Understanding the Solution**   The first ticket costs $19 and each ticket after that costs $7. Once you know that adding multiples of 7 is the rule, you can write a rule for finding the value of the nth term, $19 + (n - 1)7$.

TRY IT!

Samia is selling raffle tickets. If you buy more than one ticket, the price per ticket is lower. Use the table below to predict how much it would cost to buy 9 tickets. Write a rule for the value of the nth term.

Tickets	1	2	3	4	5
Cost ($)	2.50	3.00	3.50	4.00	4.50

EXAMPLE 2

Find the missing number in the sequence 12, 24, 48, 96, ___, 384, 768.

You can solve this problem by looking for a pattern and applying the rule.

When the numbers in a sequence increase, the pattern may be to add or multiply.

Since the given numbers increase, check if adding the same number gives you each number in the sequence. $24 - 12 = \mathbf{12}$, so the pattern may be to add 12. However, $24 + \mathbf{12} = 36$, not 48.

Adding the same number does not work, so try multiplying: $12 \times 2 = 24$, $24 \times 2 = 48$, and $48 \times 2 = 96$, so the pattern is to multiply by 2.

Use the rule to get the missing number: $96 \times 2 = 192$.

▶ **Understanding the Solution**   To check your answer, find the last two numbers in the sequence. $192 \times 2 = 384$ and $384 \times 2 = 768$, so the rule is correct.

TRY IT!

Find the missing number in the sequence 4, 20, 100, 500, ___, 12,500, 62,500.

EXAMPLE 3

Armin and Rita have a contest to see who can reach the number 5,000 on their calculator first. Both start with the number 3. After that, Armin gets each new number by squaring the previous number. Rita gets each new number by multiplying by 10. Who will reach 5,000 first?

To solve this problem, remember that when you square a number, you multiply it by itself. Begin with the number 3 and use the rule for each person's pattern.

Armin: $3^2 = 3 \times 3 = 9$, $9^2 = 9 \times 9 = 81$, $81^2 = 81 \times 81 = 6,561$.

The first four numbers in Armin's sequence are 3, 9, 81, and 6,561. The value of the nth term in this sequence is n^2.

The first four numbers in Rita's sequence are 3, 30, 300, and 3,000. Armin has reached (and gone past) 5,000 first.

▷ **Understanding the Solution**   You can find Armin's sequence by entering the number 3 on your calculator and hitting the x^2 key three times. To find Rita's sequence, you can enter 3 and multiply by 10, then multiply 30 by 10 over and over again. This works because an algebraic sequence is created by performing the same operation each time on the previous answer.

TRY IT!

A sequence starts with the number 5. Each number after 5 is the sum of the square of the previous number and 1. Use your calculator to find the next three numbers in the sequence.

Exercises

SHORT RESPONSE

1 A store is having a sale on nuts. The first pound of nuts costs $1.20, and every pound after that costs only $0.50 more than the previous pound.

a. Use the rule to complete the table to the right.

b. How much will 23 pounds of nuts cost?

Answer _____

c. How much will n pounds cost?

Answer _____

Pounds	Cost ($)
1	1.20
2	1.20 + 0.50 = 1.70
3	
4	
5	

2 Find the missing value.

72, 36, 18, ____ , 4.5

A −18

B $\frac{1}{2}$

C 8

D 9

3 Use the rule to find the next number in the sequence below.

Rule: The square root of a number gives the next number in the sequence.

625, 25, ____

F 2

G 4

H 5

J 6

4 A health club charges an initial fee of $65 to join, plus a monthly membership fee of $86. The table shows the cost of joining the health club for 5 months.

Months	Cost ($)
1	151
2	237
3	323
4	409
5	495

Use the table to predict the cost of joining the club for 1 year.

A $1,097

B $1,032

C $1,011

D $866

5 A store is selling packages of cheese that weigh different amounts. The table below gives the weight in pounds of each package and the cost of the cheese.

Weight (lbs)	Cost ($)
1.5	2.50
2	3.25
2.5	4.00
3	4.75
3.5	5.50

Use the table to predict the cost of 7.5 pounds of cheese.

F $3.50

G $11.50

H $24.50

J $38.50

6 Which algebraic expression represents the sequence 2, 6, 12, 20, … ?

A $2n$

B $n^2 - 1$

C $n(n + 1)$

D $2n^2$

7 Find the next number in the sequence.

11.01, 1.101, 0.1101, ____

F 0.11011

G 0.01101

H 0.01010

J 0.11101

EXTENDED RESPONSE

8 Find the pattern in the sequence.

____ , ____ , 16, 256, 65,536

Part A

What are the first two terms in the sequence?

Answer ________________

Part B

Describe the rule of the pattern using words.

____

____

____

9 Find the pattern in the sequence.

____ , ____ , 16, 25, 36, ...

Part A

What are the first two terms in the sequence?

Answer ________________

Part B

Write an algebraic expression that represents the sequence.

Answer ________________

LESSON 1.6 — Identifying Patterns in Polygons

New York Performance Indicators

7.A.9 Build a pattern to develop a rule for determining the sum of the interior angles of polygons

7.PS.5 Make conjectures from generalizations

7.RP.6 Support an argument by using a systemic approach to test more than one case

7.CN.5 Understand how concepts, procedures, and mathematical results in one area of mathematics can be used to solve problems in other areas of mathematics

VOCABULARY

A **polygon** is a closed plane figure that is formed by three or more line segments called sides.

A **vertex** is a point where two sides meet.

The **interior angles** of a polygon are the angles inside the polygon.

A **diagonal** of a polygon is a line segment that joins two nonconsecutive vertices.

REVIEW

Understanding Polygons

Polygons are named for the number of their sides.

Number of Sides	Polygon
3	Triangle
4	Quadrilateral
5	Pentagon
6	Hexagon
7	Heptagon
8	Octagon

What You Should Know

The measures of the interior angles of any triangle add up to 180 degrees.

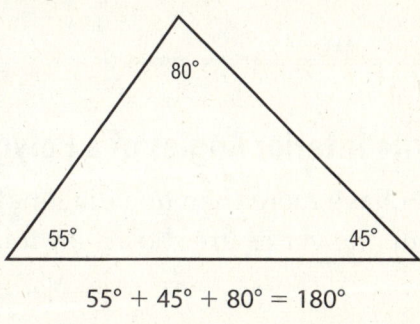

$55° + 45° + 80° = 180°$

The Sum of the Interior Angles of a Quadrilateral

Each of the four angles in a rectangle are right or 90° angles. Since $4 \times 90 = 360$, the sum of the measures of the interior angles of a rectangle is 360°. A rectangle is a special quadrilateral, but you can find the sum of the measures of the interior angles of any quadrilateral.

EXAMPLE 1

What is the sum of the measures of the interior angles of a quadrilateral?

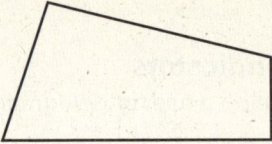

You can solve this problem by using a diagonal to divide the quadrilateral.

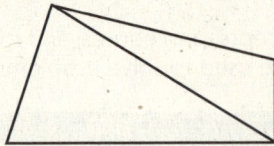

You already know that the measures of the interior angles of a triangle add up to 180°. Since the quadrilateral is made up of 2 triangles, the sum of the measures of its interior angles is 2(180°) = 360°.

▶ **Understanding the Solution**   Dividing the quadrilateral into two triangles does not change the sum of the measures of the interior angles in the quadrilateral.

TRY IT!

What is the sum of the measures of the interior angles of
a pentagon?

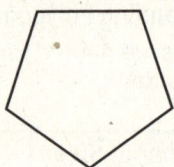

The Sum of the Interior Angles of a Polygon

A polygon can have more than one diagonal from each vertex. All possible diagonals from one vertex are shown for four polygons. A triangle has no diagonals.

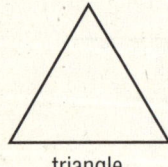   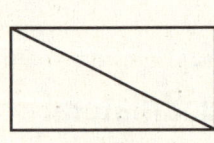   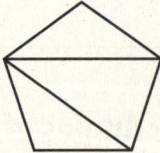   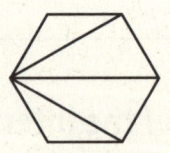

triangle   quadrilateral   pentagon   hexagon

You can summarize your results in a table to see if you can find a pattern:

Number of Sides, n	Number of Diagonals from One Vertex	Number of Triangles Formed	Sum of the Measures of the Interior Angles
3	0	1	1 × 180 = 180°
4	1	2	2 × 180 = 360°
5	2	3	3 × 180 = 540°
6	3	4	4 × 180 = 720°

Notice that the sum of the degree measures of the interior angles is the number of triangles formed, multiplied by 180°. What is the pattern between the number of sides of a polygon and the number of triangles formed? In the table, notice that the number of triangles is 2 less than the number of sides. If a polygon has n sides, the number of triangles formed is $n - 2$.

The measures of the interior angles of a polygon with n sides add up to $(n - 2)180$.

EXAMPLE 2

Find the sum of the measures of the interior angles of an octagon.

You can solve this problem by using the rule for the sum of the measures of the interior angles of a polygon.
An octagon has 8 sides, so $n = 8$.

$$(n - 2)180° = (8 - 2)180° = 6 \times 180° \text{ or } 1{,}080°$$

▶ **Understanding the Solution** Another pattern in the table is to add 180 each time. The sum of the measures of the interior angles of a hexagon is 720°. For a heptagon, the sum would be 720° + 180° = 900°. And for an octagon, the sum would be 900° + 180° = 1,080°.

TRY IT!

The sum of the measures of the interior angles in a polygon is 1,440°. How many sides does this polygon have?

Exercises

SHORT RESPONSE

1 In a regular polygon, all angles have the same measure and all sides have the same measure. Use the rule for the sum of the interior angles of a polygon to find the measure of each interior angle in a regular pentagon.

Show your work.

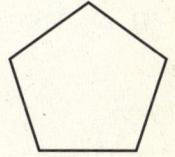

Answer _____

2 Find the sum of the measures of the interior angles of a polygon with 10 sides.

   A 1,440°

   B 1,260°

   C 1,080°

   D 900°

3 From a vertex of an octagon, all possible diagonals are drawn. How many triangles will be formed?

   F 8

   G 7

   H 6

   J 5

4 A regular polygon has all sides equal and all angles equal. The table below shows the measure of each interior angle of the first three regular polygons.

Polygon	Measure of Interior Angle
Triangle	60°
Quadrilateral	90°
Pentagon	108°

From the pattern, predict the measure of the interior angle x of a regular hexagon.

   A $x = 138°$

   B $x = 126°$

   C $x = 120°$

   D $x = 118°$

5 Use the table below to complete the algebraic equation to relate the number of sides n to the sum of the interior angles.

Number of Sides, n	Number of Triangles Formed, t	Sum of the Interior Angles
3	1	180°
4	2	360°
5	3	540°

Sum of the measures of the interior angles = _____.

   F $180°(n)$

   G $180° + n$

   H $180°(n - 1)$

   J $180°(n - 2)$

6 A regular hexagon has all sides equal and all angles equal. What is the measure of an interior angle of a regular hexagon?

   A 90°

   B 108°

   C 120°

   D 540°

7 The sum of the measures of the interior angles of a polygon is 1,800°. How many sides does this polygon have?

   F 9

   G 10

   H 11

   J 12

EXTENDED RESPONSE

8 Each interior angle of a regular polygon measures 120°.

Part A

Write an equation to find the number of sides *n* in this polygon, and solve.

Show your work.

Answer ________________

Part B

Explain how you formed the equation.

____

____

____

____

____

LESSON 1.7 Relating Functions and Equations

New York Performance Indicators

7.A.10 Write an equation to represent a function from a table of values

7.RP.7 Devise ways to verify results or use counterexamples to refute incorrect statements

7.RP.8 Apply inductive reasoning in making and supporting mathematical conjectures

7.CN.1 Understand and make connections among multiple representations of the same mathematical idea

VOCABULARY

A **function** is a relationship in which each input has exactly one output.

The **function rule** is the operations applied to the input, or the equation that represents the function.

REVIEW

Understanding Functions

A function rule can be described as an equation.

Input or *x*	Equation or Function Rule	Output or *f(x)*
10	$f(x) = x + 10$	20

You can describe this relationship as $f(10) = 20$. We read this as "*f* of 10 is 20" or "the function value at 10 is 20."

Applying Functions

A store has a special deal on T-shirts. Express the relationship between the number of T-shirts that you pay for and the number that you get as a function.

T-Shirts Bought	Total T-Shirts Received
2	4
4	8
6	12

The equation $f(x) = 2x$ shows that for every T-shirt you buy, you get one free (or, you get twice as many T-shirts as you paid for).

Finding the Function Rule from a Table

You can describe a pattern in an input/output table as a function.

EXAMPLE 1

A theater teacher thinks that the number of seniors in a school play should depend on the number of juniors in the play. The following table represents the number of juniors and seniors who performed in five different school plays.

x (Juniors)	f(x) (Seniors)
14	7
16	8
18	9
20	10

Find $f(18)$ using the table. Then find the rule that determines the relationship between 18 and $f(18)$.

You can solve this problem by thinking of 18 as the input or number of juniors, and f(18) as the output or number of seniors.

Since 18 juniors and 9 seniors are listed in the same row in the table, $f(18) = 9$. Observe the pattern in the table: the number of seniors, or $f(x)$, is half the number of juniors, or x.

So $f(18)$ is the result of plugging 18 into the function rule $f(x) = \frac{1}{2}x$.

▶ **Understanding the Solution**   You can use the function rule to find the number of seniors when the number of juniors is 18: $f(18) = \frac{18}{2} = 9$.

TRY IT!

In a restaurant, the manager observes the number of people inside and the number of cars in the parking lot at different times in the day. He comes up with the following table:

Cars Outside	People Inside
9	27
10	30
11	33
12	36

Use the table to find $f(10)$. Then find the rule that determines the relationship between 10 and $f(10)$.

EXAMPLE 2

Find whether the equation $f(x) = x + 2$ describes the function rule in the given table. If not, what is the correct equation? Use the correct equation to find the output for $x = 5$. Write your result in function notation.

You can solve this problem by comparing the output values from the equation to the output values in the table.

x	f(x)
1	1
2	3
3	5
4	7

For $x = 3, f(3) = 3 + 2 = 5$. This matches the value in the table.

For $x = 1, f(1) = 1 + 2 = 3$. This does NOT match the value in the table, so this equation is not the correct function rule.

Look at the pattern in the table to find the function value. For every input x, the output is one less than $2x$, or $f(x) = 2x - 1$. Check to see if the output values for the equation match the values in the table.

$$f(1) = 2(1) - 1 = 1$$
$$f(2) = 2(2) - 1 = 3$$
$$f(3) = 2(3) - 1 = 5$$
$$f(4) = 2(4) - 1 = 7$$

The output values match the table. Now find the output for $x = 5$ by substituting 5 for x in the function: $2x - 1 = 2(5) - 1 = 9$.
Write this in function notation: $f(5) = 9$.

▶ **Understanding the Solution**   The table shows that you can add 2 to one **output** to get the next **output**. However, this pattern does not show how to use the *input* to get the **output**. To see that pattern, you must **look across the table** to find the function rule.

TRY IT!

Find whether $f(x) = x + 3$ is the correct equation to represent the number of people in the restaurant if x represents the number of cars. Use the correct equation to predict how many people will be inside when there are 15 cars. Write your result in function notation.

Cars Outside	People Inside
9	27
10	30
11	33
12	36

Exercises

SHORT RESPONSE

1 Does the following table show a function? Explain how you know.

x	$f(x)$
16	−4
9	−3
4	−2
4	2
9	3
16	4

_____

_____

_____

_____

MULTIPLE CHOICE

2 Use the equation $f(x) = 4x + 5$ to find $f(8)$.

  A $\frac{3}{4}$

  B 8

  C 32

  D 37

3 Use the table below to find $f(4)$.

x	$f(x)$
2	2
3	4
4	6
5	8

  F 3

  G 4

  H 6

  J 8

4 Which equation shows the relationship between the input and output values in the table for question 3?

  A $f(x) = x + 2$

  B $f(x) = 2x - 2$

  C $f(x) = 2x$

  D $f(x) = x + 1$

5 Write the following in function notation: an input of 3 has an output of 17.

  F $f(3) = 17$

  G $f(17) = 3$

  H $f(3) = 3 \times 17$

  J $f(17) = 17 - 3$

6 The table below shows the sum of the interior angles for a shape with x sides.

x	$f(x)$
3	180
4	360
5	540
6	720

Choose the equation that relates the input to the output.

  A $f(x) = x + 360$

  B $f(x) = 180x$

  C $f(x) = 360(x - 1)$

  D $f(x) = 180(x - 2)$

7 The following table shows the costs for a field trip, depending on the number of students who go.

Students	Cost ($)
1	4
3	12
5	20
7	28

Choose the equation that relates the input to the output.

  F $f(x) = x + 4$

  G $f(x) = 4x$

  H $f(x) = x + 8$

  J $f(x) = x + 2$

8 Write a function rule to find all odd numbers less than 10, and complete the table.

Part A

The input values are shown in the table to the right.

x	f(x)
1	
2	
3	
4	
5	

Answer _____

Part B

Write another function rule to find all the odd numbers less than 10. The input values are shown in the table.

x	f(x)
0	
1	
2	
3	
4	

Answer _____

Problem-Solving Strategy: Working Backward

 New York Performance Indicators
8.PS.9 Work backward from a solution
8.PS.15 Choose methods for obtaining required information

Understand the Strategy

You can work backward to find an amount or quantity if you know how it has changed. Suppose you know that the height of a plant doubles every 5 days, and it is 30 inches high. You can work backward from the plant's current height to find its height from 5 days ago, 10 days ago, and so on.

Problem: A boat headed east into the Erie Canal from lock 23 at Oneida Lake. It traveled 3 hours to reach lock 22, just 15 minutes to reach lock 21, and then 110 minutes to reach lock 20. If the boat arrived at lock 20 at 3:30 P.M., at what time did it leave Oneida Lake?

SOLUTION

What do you know?

Boat starts at lock 23

From lock 23 to 22: 3 hrs

From lock 22 to 21: 15 min

From lock 21 to 20: 110 min or 1 hr, 50 min

Boat arrives at lock 20 at 3:30 P.M.

What do you need to find?

Time the boat left lock 23

Find the relationship.

Since you know the time the boat arrived at lock 20 and how long it took to travel between locks, you can work backward to find the time the boat left lock 23.

Time boat arrives at lock 20	⟶	3:30 P.M.
From lock 21 to 20	1 hr, 50 min ⟶	1:40 P.M.
From lock 22 to 21	15 min ⟶	1:25 P.M.
From lock 23 to 22	3 hrs ⟶	10:25 A.M.

The boat left lock 23 at Oneida Lake at 10:25 A.M.

▶ **Understanding the Solution**   You can quickly check your solution by rounding the travel times to whole hours and estimating the sum. Six hours before 3:30 P.M. is 9:30 A.M., which is close to the answer. A better method to check your result is to add all the travel times to 10:25 A.M. and see if you get 3:30 P.M.

1 A biologist is growing bacteria in a culture for an experiment. The number of bacteria doubles every 8 hours. At the end of 2 days, there are 6,400 bacteria in the culture. How many bacteria did the biologist start with?

Show your work.

Answer _____ bacteria

2 Kenya's dad opened a savings account for her fourteenth birthday. He put in some money. After that Kenya deposited $8 every week from her allowance. At the end of 1 year, she had $498.45. Her money had earned a total interest of $7.45 in a year. How much money did Kenya's dad put into her account?

Show your work.

Answer _____

MULTIPLE CHOICE

1 Which of the following expressions is a polynomial that does NOT have any like terms?

A $3t + 4t$

B $5y - 2y + 11$

C $6a + b + 9$

D $8x$

2 A farmer is picking the ripe apples from his orchard. The table shows how many apples he picks each day. How many apples does he pick on the sixth day?

Day	Number of Apples
1	12
2	15
3	18
4	21

F 22

G 24

H 27

J 28

3 Which expression is equivalent to the expression $9a + 7b - 5b + 2a$?

A $16a - 3b$

B $11a + 2b$

C $13ab$

D $4a + 9b$

4 The following table shows the cost of a birthday party, depending on the number of guests who attend.

Guests	Cost ($)
6	36
8	44
10	52
12	60

Choose the equation that relates the input to the output.

F $f(x) = 4x + 12$

G $f(x) = 8x + 12$

H $f(x) = 6x$

J $f(x) = x + 30$

5 What is the solution to the equation $2a + 6 = 4a + 9$?

A -3

B $-\dfrac{3}{2}$

C $-\dfrac{2}{3}$

D $\dfrac{3}{2}$

6 Find the missing number in the sequence below:

5, 10, 20, _____ , 80, 160

F 25

G 30

H 40

J 50

7 Use the pattern in the table below to find the sum of the measures of the interior angles of a hexagon and a heptagon. Write each sum in the table.

Polygon	Sum of Interior Angles
Triangle	180°
Quadrilateral	360°
Pentagon	540°
Hexagon	
Heptagon	

Show your work.

On the lines below, describe the pattern shown in the table.

_____

_____

8 Juan is playing an arcade game. The game costs $0.75 to play the first time and $0.50 each additional time. Juan has $4.75. How many times can he play the arcade game?

Write an algebraic equation that describes the situation, and identify what the variable represents. Then solve the equation and write the answer on the line below.

Equation

_____

Represented by the variable _____

Show your work.

Answer _____ times

Exploring Right Triangles

New York Performance Indicators

7.G.5 Identify the right angle, hypotenuse, and legs of a right triangle

7.G.6 Explore the relationship between the lengths of the three sides of a right triangle to develop the Pythagorean Theorem

7.PS.14 Determine information required to solve a problem

7.CM.9 Increase their use of mathematical vocabulary and language when communicating with others

VOCABULARY

A **right triangle** is a triangle that contains a 90°-angle, also called a **right angle**.

The **legs** are the sides of the triangle that are adjacent to the right angle.

The **hypotenuse** is the side opposite the right angle and the longest side in a right triangle.

REVIEW

Understanding Right Triangles

The right angle is represented by a square.

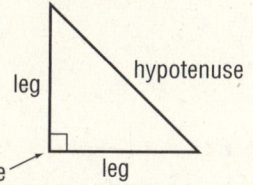

What You Should Know

In any triangle, the sum of the measures of the angles equals 180°.

Besides the right angle, the other two angles in a right triangle are acute angles, with measures less than 90°.

Exploring Right Triangles

There is a special relationship between the legs of the right triangle and the hypotenuse. You can explore this relationship by using a grid, as shown.

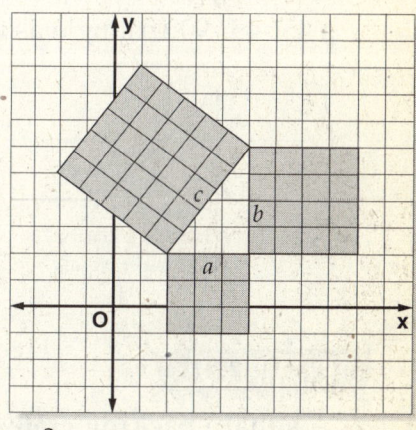

- Leg a is 3 units long.

- Leg b is 4 units long.

- Draw the hypotenuse c.

- Draw a square using leg $a = 3$. Its area is 3^2 units.

- Draw a square using leg $b = 4$. Its area is 4^2 units.

- Now draw a square along the hypotenuse. Since the hypotenuse is c units long, the area of the big square is c^2.

- All the blocks from the two smaller squares exactly cover the area of the big square. So, $c^2 = 3^2 + 4^2$.

The last equation says that the square of the length of the hypotenuse is equal to the sum of the squares of the lengths of the legs.

This relationship is true for all right triangles and is called the **Pythagorean Theorem**.

$$a^2 + b^2 = c^2$$

EXAMPLE 1

The three sides of a right triangle are 6 cm, 8 cm, and 10 cm. What is the length of the hypotenuse?

You can use the definition of a hypotenuse to solve this problem.
The hypotenuse is the longest side in a right triangle. Therefore the length of the hypotenuse must be 10 cm.

▶ **Understanding the Solution**   In any triangle, the longest side is always opposite the largest angle. The right angle is the largest angle in a right triangle, since each of the other angles must be acute, or less than 90°. So the hypotenuse is always the longest side.

TRY IT!

The three sides of a right triangle are 8 feet, 8 feet, and 11.3 feet. What is the length of the hypotenuse?

Special Right Triangles

When the two legs of a right triangle are equal, you get a special right triangle called an isosceles right triangle. The two acute angles in the triangle each measure 45 degrees, so this triangle is also called a 45°-45°-90° triangle. Using the Pythagorean Theorem, we can define the relationships between the sides of a 45°-45°-90° triangle as follows.

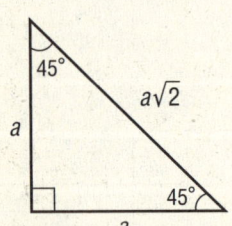

$$c^2 = a^2 + a^2$$
$$c^2 = 2a^2$$
$$c = \sqrt{2}\, a$$

The length of the hypotenuse is $\sqrt{2}$ times the length of the leg.

EXAMPLE 2

A square packing crate has a side of 10 feet and a diagonal insert to keep its contents separated. What is the length of the diagonal?

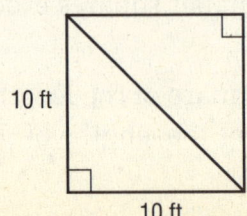

You can identify whether this is a special right triangle to help you solve the problem. You know that all the sides of a square are equal and all the corners are right angles. This means that the diagonal of the square divides it into two isosceles right triangles with legs measuring 10 feet. The diagonal is also the hypotenuse of each triangle, so you can apply the 45°-45°-90° rule to find its length.

$$c = a\sqrt{2} = 10\sqrt{2}$$

▶ **Understanding the Solution**   In a 45°-45°-90° triangle, the length of the hypotenuse c is $\sqrt{2}$ times the length of a leg a.

TRY IT!

For his school project, Jack constructed a sailboat out of geometric shapes. For the sail, he used a 45°-45°-90° triangle. The length of the hypotenuse was $9\sqrt{2}$ inches. What was the length of each leg?

Exercises

SHORT RESPONSE

1 A rope is tied from the top of a tree to the ground, as shown in the figure. The tree is perpendicular to the ground. What is the angle between the rope and the ground?

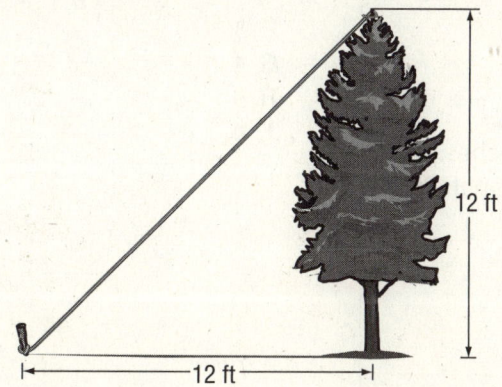

12 ft

12 ft

Answer _____

Explain how you found the angle.

2 A right triangle has legs measuring 7 inches and 24 inches. Which of these could be the length of the hypotenuse?

A 14 inches

B 17 inches

C 24 inches

D 25 inches

3 What is the length of the other leg in this triangle?

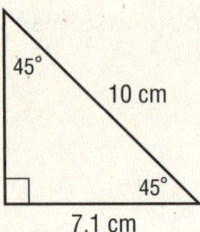

F 6 cm

G 7.1 cm

H 9 cm

J 10 cm

4 A square box has a side of 6 inches. What is the length of the diagonal of the base of the box?

A $3\sqrt{2}$ inches

B $6\sqrt{2}$ inches

C $12\sqrt{2}$ inches

D $36\sqrt{2}$ inches

5 Which of the following is true?

F An isosceles right triangle has two angles of 45 degrees.

G A right triangle has at least one obtuse angle.

H A right triangle is also an isosceles right triangle.

J The longest side of a right triangle is one of its legs.

6 The hypotenuse of a right triangle is

A next to the right angle.

B 2 times as long as the leg.

C opposite the right angle.

D equal to $\sqrt{2}$.

7 The three sides of a right triangle are related by the Pythagorean Theorem: $4^2 = 5^2 - 3^2$. What is the length of the hypotenuse?

F 3

G 4

H 5

J 9

8 Use the figure to find the length of the hypotenuse of the smaller right triangle.

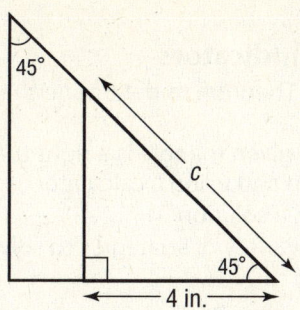

Part A

Explain how you can find the length of the missing leg.

Part B

Find the length of the hypotenuse. Explain your method below.

Answer _____ inches

Applying the Pythagorean Theorem

New York Performance Indicators

7.G.8 Use the Pythagorean Theorem to determine the unknown length of a side of a right triangle

7.G.9 Determine whether a given triangle is a right triangle by applying the Pythagorean Theorem and using a calculator

7.PS.9 Work backwards from a solution

7.CN.3 Connect and apply a variety of strategies to solve problems

---

VOCABULARY

The **Pythagorean Theorem** states that the square of the length of the hypotenuse is equal to the sum of the squares of the lengths of the legs.

REVIEW

Understanding the Pythagorean Theorem

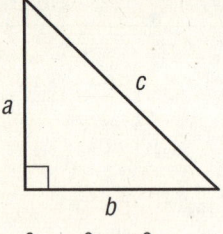

$$c^2 = a^2 + b^2$$

It is also true that if a, b, and c are the lengths of the sides of a triangle and $c^2 = a^2 + b^2$, then the triangle is a right triangle.

What You Should Know

If the length of the sides of the right triangle are whole numbers, then a, b, and c are called **Pythagorean triples**.

The multiples of a Pythagorean triple also form a Pythagorean triple.

Example
3-4-5 is a Pythagorean triple. Therefore, **6-8-10** is also a Pythagorean triple (each side was multiplied by 2).

You can use the Pythagorean Theorem to find the length of a side of a triangle or to check if a triangle is a right triangle.

EXAMPLE 1

A right triangle has legs of 9 and 12 inches. What is the length of the hypotenuse?

You can solve this problem by applying the Pythagorean Theorem.

Plug in the lengths of a and b in order to find the value of c.

$$c^2 = a^2 + b^2$$
$$c^2 = 9^2 + 12^2$$
$$c^2 = 81 + 144$$
$$c^2 = 225$$
$$c = \sqrt{225}$$
$$c = 15$$

Since length c is always positive, the negative square root is not used.

▶ **Understanding the Solution**   Besides using the Pythagorean Theorem to solve this problem, you might have recognized that 9 and 12 are multiples of 3 and 4, and part of the Pythagorean triple 9-12-15. This means the length of the hypotenuse is 15 inches.

TRY IT!

A right triangle has legs of 12 cm and 16 cm. What is the length of the hypotenuse?

EXAMPLE 2

Using right triangle ABC, find a.

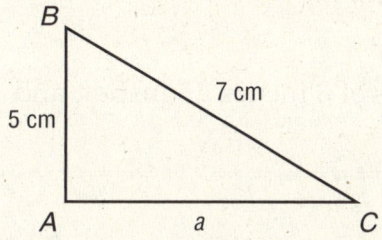

You can solve this problem by using the Pythagorean Theorem and substituting the numbers.

This time you are given the length of one leg and the hypotenuse, so be careful as to how you set up the equation.

$$a^2 + b^2 = c^2$$
$$a^2 + 5^2 = 7^2$$
$$a^2 + 25 = 49$$
$$a^2 = 49 - 25$$
$$a^2 = 24$$

Notice that 24 is not a perfect square. Use a calculator and round to the nearest tenth to find the length of side a.

$$a \approx 4.9 \text{ cm}$$

You can also express the length of the side as the radical $\sqrt{24}$, or $2\sqrt{6}$.

▶ **Understanding the Solution**   Notice that the leg a is shorter than the hypotenuse. You can check your result: $5^2 + (\sqrt{24})^2 = 25 + 24 = 49 = 7^2$.

TRY IT!

Given a right triangle with a side of 9 yards and a hypotenuse of 17 yards, find the length of the other leg.

EXAMPLE 3

Given $\triangle XYZ$ with sides that measure 7, 8, and 11, prove whether or not it is a right triangle.

You can solve this problem by using the Pythagorean Theorem.

Just plug in the numbers and see if the sides are related by the Pythagorean Theorem.

$$a^2 + b^2 = c^2$$
$$7^2 + 8^2 \overset{?}{=} 11^2$$
$$49 + 64 \overset{?}{=} 121$$
$$113 \neq 121$$

▶ **Understanding the solution**   By applying the Pythagorean Theorem, we learn that the sum of the squares of the sides does NOT equal the square of the hypotenuse. Therefore, we can conclude that this is not a right triangle.

TRY IT!

Triangle LMN has sides of 8 inches, 15 inches, and 17 inches. Is it a right triangle? How do you know?

Exercises

SHORT RESPONSE

1 Given a right triangle with legs of 7 and 9 feet, show how you would find the hypotenuse and solve to the nearest foot. Use a calculator.

Answer _____ feet

2 An apartment window is 12 feet above the ground. The top of a 15-foot ladder is resting on the windowsill. How far from the foot of the building is the bottom of the ladder?

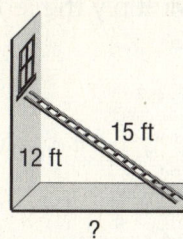

15 ft

12 ft

?

A 6 feet
B 7 feet
C 9 feet
D 10 feet

3 Using right triangle *ABC*, what is the length of the hypotenuse rounded to the nearest tenth?

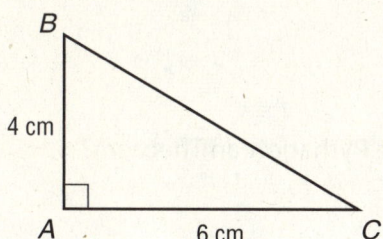

B

4 cm

A　　6 cm　　*C*

F 6.8 cm
G 7.2 cm
H 12.1 cm
J 15 cm

4 Which of the following is *not* a Pythagorean triple?

A 15, 20, 25
B 9, 12, 15
C 21, 24, 35
D 12, 16, 20

5 A square table has a diagonal of 8 feet. What is the length of each side?

F $\sqrt{2}$ feet
G $\sqrt{16}$ feet
H $\sqrt{32}$ feet
J $\sqrt{64}$ feet

6 After takeoff, an airplane is flying at a height of 8,000 feet above ground. It has traveled a distance of 10,000 feet from the airport. If Sam was to drive from the airport to the spot under the airplane, how far would he have to travel?

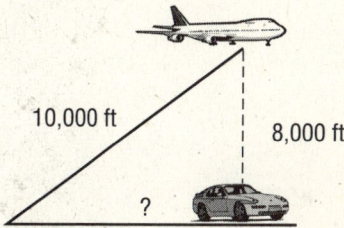

10,000 ft

8,000 ft

?

A 4,000 feet
B 6,000 feet
C 7,500 feet
D 11,500 feet

7 Is triangle *ABC* a right triangle? Why or why not?

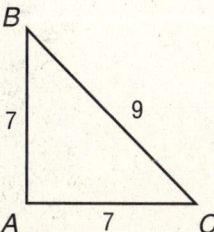

B

7　　　9

A　　7　　*C*

F yes, because it is a 45°-45°-90° triangle
G yes, because applying the Pythagorean Theorem works
H no, because it is not an isosceles triangle
J no, because the converse of the Pythagorean Theorem does not work

8 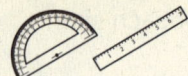 Use your protractor and the centimeter side of your ruler to help you solve this problem.

A right triangle has sides that measure 3 cm, 4 cm, and 5 cm. Multiply the length of each side by 2.

Part A

Draw the new triangle below and show the length of each side.

Are the sides of the new triangle also related by the Pythagorean Theorem?
Show your work.

Answer _____

Part B

Use the same method to find another right triangle whose hypotenuse is 15.
Explain your answer.

Answer _____

Problem-Solving Strategy: Using Information in Figures

New York Performance Indicators

8.PS.14 Determine information required to solve the problem

8.RP.2 Use mathematical strategies to reach a conclusion

Understand the Strategy

In many geometry problems with figures, part of the information you need may be in the figure. For example, if a figure is drawn in a coordinate plane, then you can find the lengths of the line segments from their coordinates.

Problem: Find the length of the hypotenuse of the right triangle shown to the right. Write the length to the nearest integer.

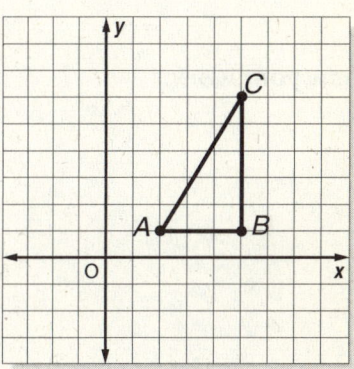

SOLUTION

What do you know?

The vertices of the right triangle:
$A(2, 1)$, $B(5, 1)$, $C(5, 6)$

What do you need to find?

The length of the hypotenuse

What is the relationship?

To find the length of the hypotenuse, you need the lengths of the two legs. Then you can use the Pythagorean Theorem.

You can find the length of the leg $\overline{AB}$ by subtracting the x-coordinates of its end points: $\longrightarrow$ $5 - 2 = 3$

You can find the length of the leg $\overline{BC}$ by subtracting the y-coordinates of its end points: $\longrightarrow$ $6 - 1 = 5$

Now use the Pythagorean Theorem to find the length of the hypotenuse:

$$AC = \sqrt{3^2 + 5^2}$$
$$= \sqrt{9 + 25}$$
$$= \sqrt{34}$$
$$\approx 5.8$$

▶ **Understanding the Solution**   When finding lengths, make sure that the answer is rounded correctly. To the nearest integer, 5.8 rounds up to 6.

Exercises

1 Find the length of the diagonal of the square shown below.

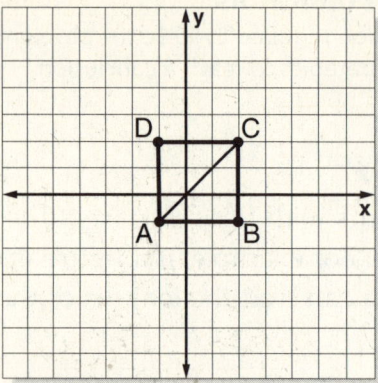

Show your work.

Answer _____ units

2 Triangle *ABC* is a 45º-45º-90º triangle. A smaller right triangle *ADE* is shown. Are the two legs of the smaller right triangle equal?

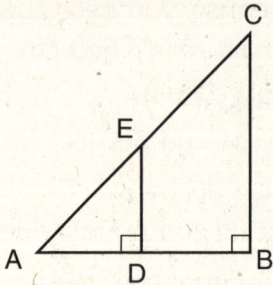

Write and explain your answer on the lines below.

MULTIPLE CHOICE

1 The two legs of a right triangle measure 12 inches and 16 inches. How long is the hypotenuse?

  A 16 inches
  B 20 inches
  C $16\sqrt{2}$ inches
  D $20\sqrt{2}$ inches

2 A square pane of glass has sides of 15 cm. How long is its diagonal?

  F 15 cm
  G $15\sqrt{2}$ cm
  H $15\sqrt{3}$ cm
  J $30\sqrt{3}$ cm

3 A leg of an isosceles right triangle is

  A half as long as the hypotenuse.
  B the same length as the other leg.
  C twice as long as the other leg.
  D twice as long as the hypotenuse.

4 A right triangle has sides of 12 km, 20 km, and 16 km. Identify the length of the hypotenuse.

  F 12 km
  G 16 km
  H $2\sqrt{20}$ km
  J 20 km

5 What is the length of the hypotenuse of right triangle *ABC*?

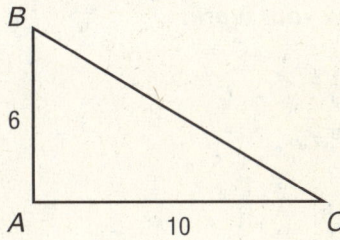

  A 11
  B $\sqrt{136}$
  C 16
  D $2\sqrt{106}$

6 A triangle has sides measuring 78 mm, 30 mm, and 72 mm. Which equation would you use to test whether it is a right triangle?

  F $78^2 + 30^2 = 72^2$
  G $\sqrt{72} + \sqrt{30} = \sqrt{78}$
  H $\sqrt{78} + \sqrt{30} = \sqrt{72}$
  J $72^2 + 30^2 = 78^2$

7 The diagonal of a square garden is 24 feet. Jane wants to build a fence around the garden. In order to buy enough fencing, Jane needs to find the perimeter of the garden.

Part A

Draw a diagram of the garden and label the known lengths.

Part B

Find the length of the side of the square garden to the nearest foot.

Show your work:

Answer _____ feet

Part C

What is the perimeter of the garden?

Show your work:

Perimeter _____ feet

Finding Distances on a Map

New York Performance Indicators

7.M.1   Calculate distances using a map scale

7.PS.10 Use proportionality to model problems

7.CN.6 Recognize and provide examples of the presence of mathematics in their daily lives

7.R.9   Use mathematics to show and understand physical phenomena (e.g., make and interpret scale drawings of figures or scale models of objects)

VOCABULARY

Similar figures have the same shape but not always the same size.

Corresponding angles and **corresponding sides** are the matching angles and sides of similar figures.

A **proportion** is an equation that sets two ratios equal.

A **scale factor** is the ratio of a length on a scale drawing or model to the corresponding, real length.

REVIEW

Understanding Proportions and Scale

The corresponding sides in similar triangles are proportional.

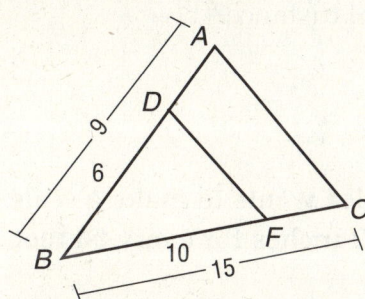

$\triangle ABC \sim \triangle DBF$, so $\dfrac{AB}{DB} = \dfrac{BC}{BF} = \dfrac{AC}{DF}$

Since, $\dfrac{BC}{BF} = \dfrac{15}{10} = 1.5$, the sides of $\triangle ABC$ are **1.5** times longer than the sides of $\triangle DBF$.

Applying Proportions and Scale

The map uses a scale of 1 cm for every 100 miles. What is the actual, straight-line distance between Buffalo and Albany in miles?

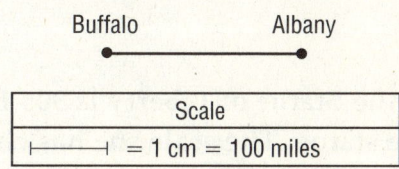

Using a ruler, you can measure the distance on the map as 2.6 cm.

$\dfrac{1}{100} = \dfrac{2.6}{x} \rightarrow x = (100)(2.6) = 260$

The cities are about 260 miles apart.

Using Scale Drawings

A scale drawing is a drawing that has been enlarged or made smaller. The measurements used in a scale drawing are proportional to the actual or real measurements. Maps are examples of scale drawings.

EXAMPLE 1

 Use your ruler to help you solve this problem.

Ithaca   Binghamton

What is the actual distance between the two cities shown on the map?

Scale
⊢————————⊣ = 1 in. = 25 miles

You can solve this problem using a proportion.

Write a ratio to represent the map scale of inches to miles. $\frac{1}{25}$

Measure the distance on the map with a ruler. The cities are about 1.6 inches apart.

Let x represent the unknown distance in miles. Then write a ratio comparing 1.6 inches to the unknown distance in miles. $\frac{1.6}{x}$

Now write a proportion using the two ratios. Find the cross-products and solve for x.

$$\frac{1}{25} = \frac{1.6}{x} \rightarrow x(1) = (25)(1.6) \rightarrow x = 40 \text{ miles}$$

▶ **Understanding the Solution** The 1.6 inches on the map represent 40 miles. Notice that you are multiplying the number of inches times the scale, or the number of miles represented by one inch.

TRY IT!

A map scale shows 2 inches for every 150 miles. If the distance between city A and city B on the map is 1.5 inches, what is the actual distance?

EXAMPLE 2

The height of the Statue of Liberty is 305 feet. Jessica wants to make a scale drawing of the statue. The scale she has chosen is 2 inches for every 50 feet. About how tall will Jessica's drawing be?

You can solve this problem using proportions.

Write a ratio for Jessica's scale. $\frac{2}{50}$

Write a ratio for the unknown length, x, over the actual length of the statue.

Write the proportion and solve for x.

$$\frac{2}{50} = \frac{x}{305} \rightarrow 50x = (2)(305) \rightarrow x = \frac{610}{50} = 12.2 \text{ inches}$$

▶ **Understanding the Solution** The important thing is to make sure that both ratios in the proportion compare units in the same order: $\frac{2 \text{ in.}}{50 \text{ ft}} = \frac{x \text{ in.}}{305 \text{ ft}}$.

TRY IT!

Jaime wants to make a scale model of a tree in his backyard for a school project. The tree is 24 feet tall. Jaime wants to use a scale of 0.5 inches for every foot. How tall will his model be?

Using Similar Figures

Since you know that corresponding sides of similar figures are proportional, you can find the unknown length of a side using a proportion.

EXAMPLE 3

A tree's shadow is 10 feet long. A 5-foot pole next to it casts a shadow that is 4 feet long. If the triangles formed by each object and its shadow are similar, how tall is the tree?

You can solve this problem using a proportion with ratios that compare similar sides.

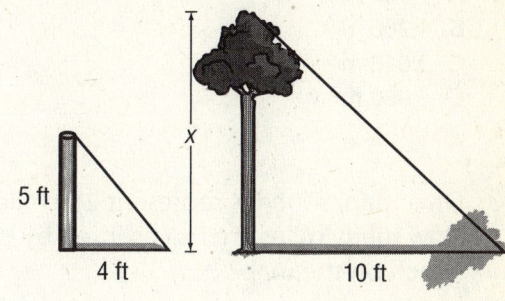

$$\frac{5}{x} = \frac{4}{10} \rightarrow 4x = 50 \rightarrow x = \frac{50}{4} = 12.5 \text{ feet}$$

▶ **Understanding the Solution** The first ratio, $\frac{5}{x}$, compares the known height of the pole to the unknown height of the tree. The ratio $\frac{4}{10}$ compares the lengths of the shadows, or the bases in the triangles, in the same order.

TRY IT!

Find the length of the missing side using a proportion for the similar triangles shown.

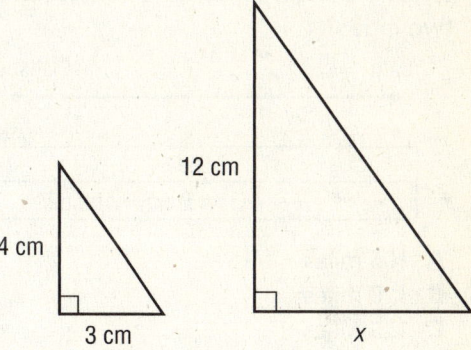

Exercises

Short Response

1 What value of *h* would make the triangles similar?

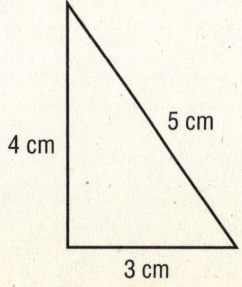

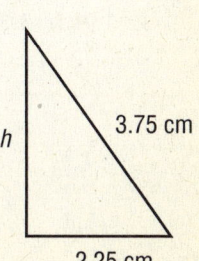

Answer _____ cm

2 A drawing of a suspension bridge uses a scale of 1 cm = 200 m. In the scale drawing, one span of the bridge is 6 cm long. What is the actual length of that span of the bridge in meters?

A 1,000 m

B 1,200 m

C 3,000 m

D 6,000 m

3 On a map, 4 inches represent 240 miles. How many miles are represented by 1 inch on the map?

F 240 miles

G 120 miles

H 80 miles

J 60 miles

4 Use your ruler to help you solve this problem.

What is the actual distance between the two cities?

Bath   Plattsburgh

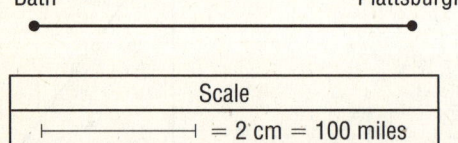

| Scale |
| ⊢————⊣ = 2 cm = 100 miles |

A 105 miles

B 120 miles

C 125 miles

D 250 miles

5 Angelo drew a scale model of the flagpole at school. If the actual pole is 15 feet tall, and his drawing is 10 inches tall, what scale did he use?

F 1 inch : 1 foot

G 1.5 inches : 1 foot

H 1.5 inches : 1.5 feet

J 1 inch : 1.5 feet

6 Dave is working on a project using similar figures. If he wants to draw two similar rectangles like the ones below, how long does the second rectangle have to be?

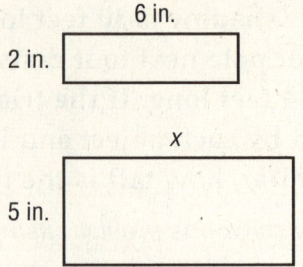

A 10 inches

B 12 inches

C 15 inches

D 20 inches

7 Karen wants to fence off a section of her yard for her puppies. The part she can fence is in the shape of the triangle below.

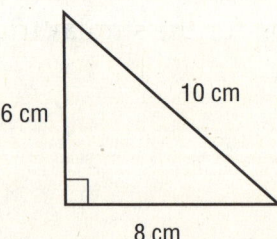

If every 2 cm on the scale drawing represents 1 m of Karen's yard, how much fencing will she need?

F 12 meters

G 24 meters

H 48 meters

J 200 meters

EXTENDED RESPONSE

8 Larry is planning a trip from Rochester to Syracuse and from Syracuse to Rochester.

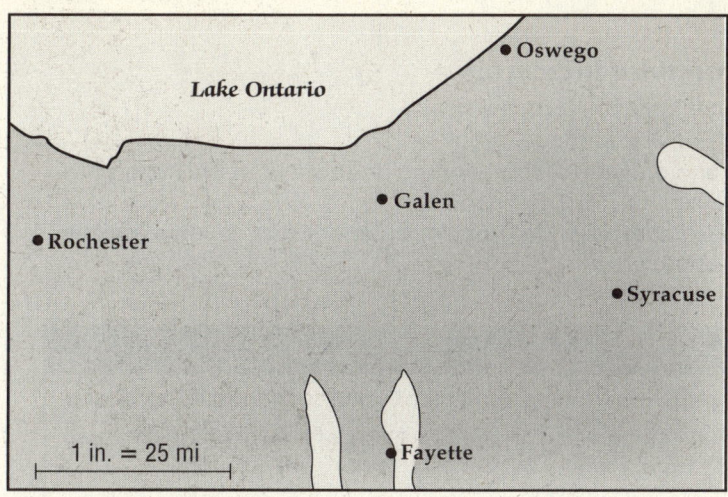

Part A

How can you find the distance between the two cities? Explain.

_____

_____

_____

Part B

Use your ruler to help you solve this problem.

How many miles does Larry need to drive to complete his trip?

Show your work.

Answer _____ miles

LESSON 3.2 Finding Unit Prices

New York Performance Indicators

7.M.5 Calculate unit price using proportions

7.PS.10 Use proportionality to model problems.

7.CN.6 Recognize and provide examples of the presence of mathematics in their daily lives

7.CN.7 Apply mathematical ideas to problem situations that develop outside of mathematics

VOCABULARY

A **rate** is a ratio that compares amounts in different units, such as miles to hours.

A **unit rate** is a ratio that compares an amount to 1 unit of another amount, such as distance covered in 1 hour.

A **unit price** is a unit rate that gives the cost of one item or measured amount, like dollars per pound ($/lb).

REVIEW

Understanding Unit Prices

Suppose a pound of butter costs $4.60. One pound is the same as 16 ounces.

The unit price **per pound** is $4.60.

The unit price **per stick** is $4.60/4, or $1.15.

The unit price **per ounce** is $4.60/16, or about $0.29.

What You Should Know

You can find unit rates by writing a ratio.

Once you know the unit price, you can find the cost of any amount. For example, to find how much 3 pounds of butter cost, multiply the unit price per pound times 3.

$$\$4.60 \times 3 = \$13.80$$

Finding Unit Prices

You can use ratios to find unit rates such as unit prices, speed, and gas mileage. Since unit prices involve money amounts, round decimal numbers to the nearest hundredth or nearest cent.

EXAMPLE 1

A cab ride from JFK airport to midtown Manhattan costs about $45. If the distance is about 15 miles, how much does the passenger pay for every mile?

You can solve this problem by using proportions.

Let c be the cost for 1 mile. Write a ratio for the total cost to the cost c for 1 mile. $\frac{45}{c}$

Write a ratio for the cost of 15 miles. $\frac{\$45}{15 \text{ miles}}$

Set the ratios equal in a proportion. $\frac{\$45}{15 \text{ miles}} = \frac{\$c}{1 \text{ mile}}$

Now find the cross-products and solve for c.

$$15c = 45$$
$$c = \frac{45}{15}$$
$$c = 3$$

The passenger pays $3 for every mile. Notice that you can also find the unit cost simply by dividing the total cost by the number of miles:
$45 ÷ 15 miles = $3/mile.

▶ **Understanding the Solution** Why can you skip the step for finding cross-products? The ratio $\frac{c}{1}$ represents the price for one mile. Since the unit price is a rate with a denominator of 1, the proportion is understood. Just remember to include the units in your answer.

TRY IT!

Berto paid $2.25 to ride the bus 12 miles to school. How much did he pay per mile? Remember to round your price to the nearest cent.

Using Unit Prices to Find the Cost of More Items

EXAMPLE 2

Greg paid $34.40 for 12.5 gallons of gas at the gas station. How much would he have paid for 8 gallons of gas?

You can solve this problem by finding and using the unit price of the gas.

First divide to find the unit price or cost of 1 gallon of gas.

$$\frac{34.40}{12.5} = 2.752 \text{ or about } \$2.75/\text{gal}$$

Multiply the unit price per gallon times 8 to find the cost of 8 gallons of gas.

$$8 \times \$2.75 = \$22$$

▶ **Understanding the Solution** You can also solve this problem by writing a proportion using p for the unknown price of 8 gallons: $\frac{\text{cost}}{\text{gallon}} = \frac{\$34.40}{12.5 \text{ gal}} = \frac{p}{8}$.

By cross-multiplying, $12.5p = (34.40)(8) = 275.20$ and $p = 275.20 ÷ 12.5 = 22.016$, or about $22.

TRY IT!

Vicki paid $21.45 for 24 juice bottles. How much would she have paid for 16 juice bottles?

Exercises

1 The table below shows the price of paper towels bought at a warehouse store. Use proportions to find the unit price for each and find which is the better buy. Round to the nearest cent.

Super Club Prices

Super Absorbent Towels	9 rolls for $5.99
Thirsty Power Towels	12 rolls for $7.29

Show your work.

Super Absorbent _____

Thirsty Power _____

Answer _____ Towels are the better buy.

2 Michael argued that paying $0.89 for 1 doughnut is the same as paying $1.89 for 2 doughnuts because you add 1 unit to each of the quantities you compare in the ratio. Is he correct? Explain.

_____

_____

_____

_____

3 A computer technician charged $90 for 2 and one-half hours of work. What is his hourly rate?

   A $30 per hour

   B $36 per hour

   C $40 per hour

   D $45 per hour

4 Jeans are on sale at the Jean Depot. You can buy two pairs of jeans for $42.98. If you buy the two pairs, what did you pay for each pair?

   F $20.00

   G $20.49

   H $21.49

   J $42.98

5 Raul's dad filled his car with gas. He paid $44.85 for 15 gallons of gas. What is the unit price for the gasoline?

   A $2.42/gallon

   B $2.99/gallon

   C $4.48/gallon

   D $4.85/gallon

6 Tina is selling greeting cards with her own artwork on them. She can sell a box of 9 cards for $12.00. She wants to sell individual cards for about 50 cents more per card than if someone bought the box. What would she charge for individual cards, rounded to the nearest cent?

   F $0.50 per card

   G $1.33 per card

   H $1.50 per card

   J $1.83 per card

7 It costs a furniture maker $189 to buy wood and other materials and make 10 wooden chairs. Each chair is sold for $42. How much profit does the furniture maker make per chair?

   A $23.10/chair

   B $23.90/chair

   C $42.00/chair

   D $60.90/chair

8 Amanda has $5.00. She buys 5 bags of chips for herself and four friends. If she gets $0.55 back in change, how much does each bag of chips cost?

   F $0.45

   G $0.55

   H $0.89

   J $1.00

9 Danny goes to a party store to get birthday hats. A sign says 10 hats for $3.99. Danny needs 16 hats for the party. He has $6 to spend on hats. Does Danny have enough money to buy the hats? Solve this problem using two different methods.

Show your work.

Method 1

Method 2

Answer _____

LESSON 3.3 Comparing Unit Prices

New York Performance Indicators

7.M.6 Compare unit prices

7.PS.2 Construct appropriate extensions to problem situations

7.PS.10 Use proportionality to model problems

7.CN.6 Recognize and provide examples of the presence of mathematics in their daily lives

REVIEW

Understanding Comparisons of Unit Prices

Look at the two bottles of juice.

Amount	Total Price	Unit Price per oz
16 oz	$1.99	$0.12
40 oz	$3.29	$0.08

$0.12 > $0.08, so the larger bottle costs less **per ounce**.

Applying Comparisons of Unit Prices

You can buy a box of 12 doughnuts for $3.99, or buy 12 single doughnuts for $0.45 each. Which is the better buy?

box of 12 doughnuts:

$$\frac{\$3.99}{12 \text{ doughnuts}} \approx \$0.33 \text{ per doughnut}$$

$0.33 < $0.45, so it is cheaper to buy 12 doughnuts in a box than 12 single doughnuts. The box of doughnuts is the better buy.

You can use unit prices to compare costs and decide which item is the better buy. Make sure to round each unit price to the nearest cent.

Comparing Unit Prices

EXAMPLE 1

The table below shows the number of minutes you can purchase with different calling cards. Which calling card is the better buy?

Calling Card	Number of Minutes	Price ($)
CALLMOM	30 minutes	2.50
FONE	45 minutes	4.00

You can solve this problem by finding and comparing the unit price of each card.

$$\text{CALLMOM} \quad \frac{\$2.50}{30 \text{ min}} = \$2.50 \div 30 \approx \$0.08 \text{ per minute}$$

$$\text{FONE} \quad \frac{\$4.00}{45 \text{ min}} = \$4.00 \div 45 \approx \$0.09 \text{ per minute}$$

$$\$0.08 < \$0.09, \text{ so } \textbf{CALLMOM} \text{ is the better buy.}$$

▶ **Understanding the Solution** The better buy has a lower unit price. Rounded to the nearest cent, the unit price for the CALLMOM card is less than the unit price for the FONE card.

TRY IT!

A 20-ounce box of graham crackers costs $3.29. A 16-ounce box costs $2.89. Find the unit price for each. Which is the better buy?

EXAMPLE 2

The unit price of a T-shirt is $6.50 when purchased in a pack of 3. A single T-shirt costs $8. What is the least amount of money Sandy can spend to get 4 T-shirts?

You can solve this problem by using the unit price.

To spend the least amount of money, Sandy can buy a 3-pack and a single T-shirt.

$$\text{Cost of the 3-pack} = \$6.50 \times 3 = \$19.50$$
$$\text{Cost of a single shirt} = \$8$$
$$\text{Total cost} = \$19.50 + \$8 = \$27.50$$

▶ **Understanding the Solution** Buying 4 T-shirts individually would have cost $4 \times 8 = \$32$.

TRY IT!

The unit cost of a juice bottle is $0.69 when purchased in a pack of 6. A single juice bottle costs $0.85. Eugenia wants to buy 9 juice bottles. What is the least amount of money she can spend for 9 bottles?

Exercises

1 The table below shows the price of paper plates at the grocery store. Find the unit price for each brand, rounded to the nearest cent. Which is the best buy?

Super Club Prices

Sturdy Plates	24 pack for $3.29
No Spill Plates	10 pack for $1.79
Compartment Plates	50 pack for $6.00

Show your work.

Answer _____

2 Four presents cost $28.95. What is the unit price rounded to the nearest cent?

   A $7.23 per present

   B $7.24 per present

   C $115.80 per present

   D $116.00 per present

3 Annie paid $54.50 for 2 pairs of jeans. What would 3 pairs of jeans at the same price cost?

   F $27.25

   G $54.50

   H $81.50

   J $81.75

4 The table below shows the cost of fish at the local markets. Which market offers the best buy?

Salmon on Sale

Captain's	2 lbs for $7.98
Seacatch	5 lbs for $9.75
Ocean	3 lbs for $8.25
Crab Shack	$2.99/lb

   A Captain's

   B Seacatch

   C Ocean

   D Crab Shack

5 Which of these is a unit price?

   F $\dfrac{\$2.00}{\text{gal}}$

   G $\dfrac{\text{gal}}{\$2.99}$

   H $\dfrac{\$7.50}{2 \text{ oz}}$

   J $\dfrac{2 \text{ pairs}}{\$1.00}$

6 Which is the best buy?

   A $6.50 for 6 bagels

   B $2.00 for 3 bagels

   C $7.50 for 8 bagels

   D $4.00 for 4 bagels

7 The cost for 2 pounds of apples is $2.90. At the same rate, which of the following would be true?

   F 1 pound costs $1.50.

   G 3 pounds cost $4.35.

   H 4 pounds cost $5.95.

   J 5 pounds cost $6.50.

8 Fariz goes to buy laundry detergent at a local store.

A 32-oz box of Super Clean costs $5.50.

A 40-oz box of Tough Cycle costs 10% more than the 32-oz box of Super Clean.

Part A

How much does a 40-oz box of Tough Cycle cost?

Show your work.

Answer _____

Part B

Which box of detergent is a better buy?

Show your work.

Answer _____ is a better buy.

LESSON 3.4 Converting Money

 New York Performance Indicators

7.M.7 Convert money between different currencies with the use of an exchange rate table and a calculator

7.PS.15 Choose methods for obtaining required information

7.R.10 Use mathematics to show and understand social phenomena (e.g., determine profit from sale of yearbooks)

VOCABULARY

Currency is paper money. The unit of currency in the U.S. is the dollar ($).

An **exchange rate** is a rate that compares one unit of the currency of one country to the currency of another country.

REVIEW

Understanding Currency Conversion

You can convert back and forth between different currencies. For example, say the exchange rate between Mexican pesos and U.S. dollars is 11.371 pesos per $1.

For $30 you can get 341.13 pesos:

$$\$30 \times \frac{11.371 \text{ pesos}}{\$1} = 341.13 \text{ pesos}$$

To convert U.S. dollars to another currency, **multiply** by the exchange rate per $1.

What You Should Know

To convert from another currency to U.S. dollars, **divide** by the exchange rate per $1, or multiply by the reciprocal.

If you have 30 pesos, you can get approximately $2.64:

$$30 \text{ pesos} \times \frac{\$1}{11.371 \text{ pesos}} \approx \$2.64$$

Converting Between U.S. Dollars and Other Currencies

The exchange rate can change from day to day. The table shows some average rates for the month of June 2005.

Average Monthly Exchange Rate for June

Country	Monetary Unit	Rate per U.S. Dollar
Brazil	Real	2.4143
Canada	Dollar	1.2403
China, P.R.	Yuan	8.2765
European Union	Euro	1.2154
India	Rupee	43.5200
Japan	Yen	108.7500
Mexico	Peso	10.8190
South Africa	Rand	6.7396
United Kingdom	Pound	1.8177
Venezuela	Bolivar	2144.6000

EXAMPLE 1

Samia travels from Brazil to the United States and exchanges 230 reals for U.S. dollars. Use the appropriate exchange rate from the table on the previous page to find the number of dollars she will get.

You can solve this problem using a proportion.

One dollar is worth 2.4143 reals. Let x = the number of dollars worth 230 reals. Then set up the following proportion:

$$\frac{2.4143 \text{ reals}}{1 \text{ dollar}} = \frac{230 \text{ reals}}{x \text{ dollars}}$$

Find the cross-product and solve for x. $2.4143x = 230 \rightarrow x = \frac{230}{2.4143} \approx 95.27$

Samia can get approximately $95.27.

▶ **Understanding the Solution** Samia should end up with **fewer** dollars than reals, since there are many reals to a dollar. As long as you compare dollars and reals in the same order in each ratio, you can use a different proportion to get the same answer: $\frac{1 \text{ dollar}}{2.4131 \text{ reals}} = \frac{x \text{ dollars}}{230 \text{ reals}}$.

TRY IT!

When Shivam traveled to India from the United States, 1 U.S. dollar was worth 43.52 rupees. How many rupees did he receive when he exchanged $400?

Converting Between Foreign Currencies

EXAMPLE 2

Kerry travels from France, which uses the euro, to Mexico, which uses pesos. How many pesos can he get for 120 euros? Use the appropriate exchange rate below.

Pesos	Euros
1	0.1143
8.749	1

*You can solve this problem by **multiplying** the number of pesos you can get for **one** euro by 120 euros.*

One euro is worth 8.749 pesos. $8.749 \times 120 = 1{,}049.88$, so Kerry will get 1,049.88 pesos.

You can also solve this problem by setting up and solving the following proportion. Let x = the number of pesos worth 120 euros.

$$\frac{8.749 \text{ pesos}}{1 \text{ euro}} = \frac{x \text{ pesos}}{120 \text{ euros}} \rightarrow 8.749 \times 120 = x \rightarrow x = 1{,}049.88$$

▶ **Understanding the Solution** You can also use the other exchange rate, 0.114 euros to 1 peso, as long as both ratios compare pesos and euros in the same order:

$$\frac{1 \text{ peso}}{0.1143 \text{ euros}} = \frac{x \text{ pesos}}{120 \text{ euros}} \rightarrow 120 = 0.1143x \rightarrow x = \frac{120}{0.1143} \approx 1{,}049.87.$$

Notice that there is a slight difference in your answer due to rounding off the cent amount.

TRY IT!

If Darshan traveled from India to the United Kingdom, how many pounds would he get for 620 Rupees?

Rupees	Pounds
24.20	1
1	0.0413

Exercises

SHORT RESPONSE

1 After a summer trip to Mexico and Venezuela, Felix returns home with the following currency:

Currency	Exchange Rate per $1
251 pesos	11.371
9,000 bolivars	2144.6

How much money does Felix get after exchanging his foreign currency to U.S. dollars?
Show your work.

Answer _____

2 If the exchange rate between Venezuelan and American currencies is 2144.6 bolivars per U.S. dollar, about how many U.S. dollars can you get for 500 bolivars?

A $0.23

B $4.28

C $4.29

D $1,072,300

3 Raul is visiting London with his family. One ticket to ride the subway train from Heathrow Airport to downtown London costs 3.80 pounds, or £3.80. If the exchange rate on that day is $1.8607 per £1, how much does one ticket cost in U.S. dollars?

F $0.49

G $2.04

H $5.73

J $7.07

4 On July 1, 2005, there were 8.2765 Chinese yuan per $1 and 1.1957 euros per $1. On the same day, how many euros were equal to 1 yuan?

A 0.1445 euros

B 6.9219 euros

C 9.8962 euros

D 19.7924 euros

5 In 2005, admission to the Bronx zoo was $6 for a child and admission to the Toronto zoo was 11 Canadian dollars. Using the exchange rate of 1.2403 Canadian dollars per U.S. dollar, which zoo's admission costs more in U.S. dollars?

F Bronx zoo

G Toronto zoo

H Both zoos cost about the same.

J There is not enough information to tell.

6 Julia had a meal at a restaurant in Tokyo, Japan, that cost 1,500 yen. She ate a similar meal in New York City that cost $24.00. Using the exchange rate of 111.568 yen per U.S. dollar, in which city did Julia pay more for her meal?

A New York

B Tokyo

C Both restaurants charged about the same amount.

D There is not enough information to tell.

7 Use the table below to find the difference in the worth of 200 U.S. dollars in Sudan in January and February of 2004.

Sudanese Dinars per U.S. Dollars

January	258.26
February	259.34

F 216 dinars

G 51,652 dinars

H 51,868 dinars

J 103,520 dinars

8 The graph below shows the average monthly exchange rates of Sri Lankan rupees per U. S. dollar in 2004.

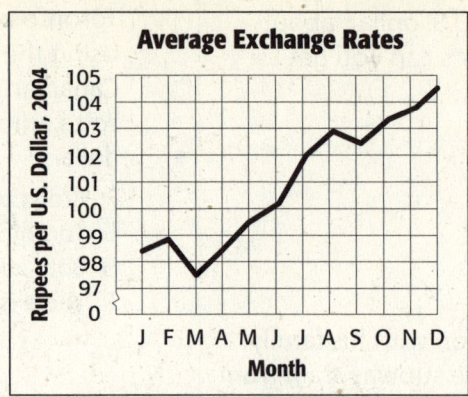

Part A

Write the names of the months that had the highest and lowest average exchange rates below. Then use the exchange rate for each month to convert $100 to rupees.

Show your work.

Lowest exchange rate _____

$100 equaled _____ rupees.

Highest exchange rate _____

$100 equaled _____ rupees.

Part B

Given these exchange rates from U.S. dollars to Sri Lankan rupees, when was the U.S. dollar the strongest in 2004? Explain your answer.

_____

_____

_____

_____

LESSON 3.5 | *Problem-Solving Strategy:*
Solving a Simpler Problem

New York Performance Indicators

8.PS.8 Understand how to break a complex problem into simpler parts or use a similar problem to solve a problem

8.PS.12 Interpret solutions within the given constraints of a problem

8.CN.3 Connect and apply a variety of strategies to solve problems

Understand the Strategy

If a problem seems difficult or complex, then you should try to simplify the problem. How? That depends on the problem. Here are a few tips:

- Break the problem into smaller steps.

- Use smaller or fewer numbers to find a relationship.

- Use a related problem that you know how to solve.

Problem: A printing company receives an order to print 16,000 posters in 3 days. The shop has 3 printers and each printer can print 450 posters in 4 hours. The company runs two 8-hour shifts per day: 8 A.M.–4 P.M. and 4 P.M.–12 A.M. The order needs to ship before 6 P.M. on the third day. Should the company accept the order?

SOLUTION

What do you know?

Posters printed by 1 machine in 4 hours = 450

Number of printers = 3

Number of 8-hour shifts per day = 2

What do you need to find?

Can the company deliver 16,000 posters in 3 days?

Find the relationship.

There is a lot of information. By keeping in mind what you have to find, you can break the problem into smaller steps.

Find out how many posters can be printed by the three machines in 1 shift.

In 4 hours, one printer can print 450 posters. $\longrightarrow$ In an 8-hour shift, the machine can print $2 \times 450 = 900$ posters.

In one shift, 3 printers can print: $\longrightarrow$ $3 \times 900 = 2700$ posters

You can organize the information in a table to find out how many total shifts can be used to print the order.

	Day 1	Day 2	Day 3	
Shift 1	2,700	2,700	2,700	
Shift 2	2,700	2,700		
Total	5,400	5,400	2,700	**13,500**

Altogether, 5,400 + 5,400 + 2,700 = 13,500 posters can be printed.

▶ **Understanding the Solution**   Since the printer can only print 13,500 of the 16,000 posters needed in 3 days, the printer should not accept the job. Another way to find the total number of posters that can be printed in 3 days is to multiply 2,700 by 5.

SHORT RESPONSE

1 Four people can make 20 bagels in one hour. At the same rate, how many bagels can 6 people make in 3 hours?

Show your work.

Answer _____ bagels

2 Ellen is moving from Yonkers to Westfield. She needs to rent a van for 2 days to bring her things 428 miles to her new home and drive 428 miles back to return the van. Which mover should Ellen use if she wants to pay the least amount of money possible?

	Cost per Day ($)	Cost per Mile ($)	Number of Free Miles
Mover A	50	0.45	0
Mover B	75	0.55	50/day
Mover C	300/day all included		

Show your work.

Answer _____

MULTIPLE CHOICE

1 Robin has $3.00. She buys 7 cookies for herself and her friends. If she gets $0.55 back in change, how much did each cookie cost?

A $2.80

B $0.75

C $0.40

D $0.35

2 If the exchange rate between Mexican pesos and U.S. dollars is 10.7 pesos per dollar, about how many dollars can you get for 300 pesos?

F $28.04

G $32.10

H $35.67

J $36.67

3 The table below shows the price of a cereal at different supermarkets. Which supermarket has the best buy?

Market A	2 boxes for $3.96
Market B	1 box for $1.93
Market C	4 boxes for $8.00
Market D	5 boxes for $9.80

A Market A

B Market B

C Market C

D Market D

4 A map of a national forest uses a scale of 1 cm = 1.5 km. Harold wants to follow a trail that is 7.5-cm long on the map. How long is the actual trail?

F 6 km

G 7.5 km

H 11.25 km

J 12.75 km

5 It costs a painter $126 to buy the paint and other materials to make 9 paintings. She sells each painting for $75. How much profit does she make per painting?

A $14

B $51

C $61

D $14

6 In June, one U.S. dollar was worth 1.24 Canadian dollars. One U.S. dollar was also worth 43.52 Indian rupees. How many rupees would it take to make one Canadian dollar?

F 0.028 rupees

G 35.10 rupees

H 42.28 rupees

J 53.96 rupees

7 Dmitri is thinking about buying a ranch. He has found ranches he likes in three different countries: South Africa, the United States, and Mexico.

Country	Size of Ranch	Price of Ranch	Exchange Rate per US$
South Africa	500 acres	1,718,000 rand	6.7396 rand
United States	350 acres	190,700 dollars	1 dollars
Mexico	850 acres	4,820,000 pesos	10.8190 pesos

Part A

What does each ranch cost per acre in the local currency? Round to the nearest hundredth.

Show your work.

South African ranch: _____ rand per acre

United States ranch: _____ dollars per acre

Mexican ranch: _____ pesos per acre

Part B

Use the exchange rates in the table above to change the unit prices you found into U.S. dollars. Round to the nearest hundredth. Which ranch is the best buy?

Show your work.

Answer _____

PART 1

1 The formula for the rate of speed R is distance D divided by time T $\left(R = \frac{D}{T}\right)$.
Jessica walks at a rate of 5 miles per hour for 4 hours. What distance does Jessica travel?

  A 20 miles

  B 5 miles

  C 4 miles

  D $\frac{5}{4}$ miles

2 According to the pattern in the following table, what would be the fifth term?

Term	Number
1	52
2	64
3	76
4	?
5	?

  F 88

  G 98

  H 100

  J 112

3 In the month of July, one U.S. dollar was worth 43.52 Indian rupees. One U.S. dollar was also worth 108.75 Japanese yen. How many yen could you exchange for one rupee in July?

  A 0.3954 yen

  B 2.4989 yen

  C 3.6582 yen

  D 152.27 yen

4 Harold takes a hike in the forest. He walks 12 km east, then makes a 90°-turn and walks 5 km south. How far is Harold from where he started his hike?

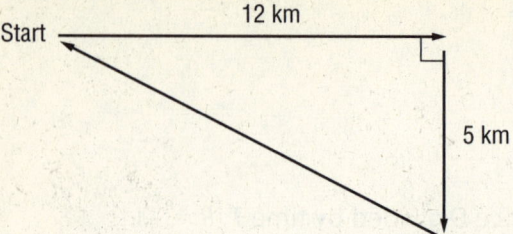

F 11 km

G 13 km

H 15 km

J 17 km

5 A rectangle is 4 inches wide and $(2z + 3)$ inches long. If the area of the rectangle is 52 square inches, what is the value of z?

```
4 in. [                                    ]
              (2z + 3) in.
```

A 3

B 4

C 5

D 6

6 Which equation represents the function values shown in the table?

x	y
−1	−5
0	−3
1	−1
2	1

F $y = x - 4$

G $y = -2x + 1$

H $y = -x - 6$

J $y = 2x - 3$

7 A clothing store is having a spring sale for a week. The table shows how the price of a shirt changes over the period of the sale.

Day of Sale	Dollars off Regular Price ($)
1	4
3	10
5	16
7	22

Which equation relates the day of the sale to the amount of the discount?

A $f(x) = x + 3$

B $f(x) = 3x + 1$

C $f(x) = 4x - 2$

D $f(x) = 2x^2$

8 The two legs of a right triangle measure 11 m and 11 m. Which expression would give you the length of the hypotenuse?

F $\sqrt{22}$

G $\sqrt{121}$

H $\sqrt{11^2 + 11^2}$

J $\sqrt{11}$

9 Which of the following deals gives you the best price per orange?

A $0.69 for 3 oranges

B $1.20 for 5 oranges

C $1.76 for 8 oranges

D $2.10 for 10 oranges

10 Sally rents a car for a day. The car costs $65 per day, plus $0.50 per mile driven. If Sally drives *m* miles in one day, which expression represents the total amount of money Sally pays for the car?

F $65 + 0.50$

G $65m + 0.50$

H $65 + 0.50m$

J $65m$

11 A weaver is weaving a rug. She begins by weaving 12 rows in one color, and then switches to another color. Every time she switches colors, she adds 14 more rows. If she switches colors 8 times, how many rows will her rug have?

A 112

B 120

C 124

D 126

12 Which equation represents the function values shown in the table?

x	y
0	−2
1	5
2	12
3	19
4	26

F $y = 5x - 2$

G $x = 7y - 2$

H $y + 2 = 7x$

J $y = 6x + 2$

13 The following table shows the population of Springville for each year y since the first year it was settled.

Year	Population
1	22,800
2	34,200
3	45,600
4	57,000

Which equation relates the input to the output?

A $f(y) = 11,400 + y$

B $f(y) = 22,800 + y$

C $f(y) = 11,400\,(y + 1)$

D $f(y) = 11,400y$

14 A rectangle is 10 cm wide and 24 cm long. How long is the diagonal of the rectangle?

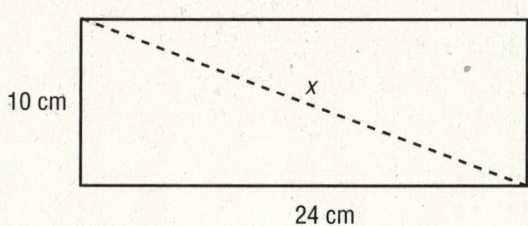

10 cm

x

24 cm

F 26 cm

G 28 cm

H 30 cm

J 34 cm

15 A right triangle has one leg measuring 15 meters and a hypotenuse measuring 25 meters. How long must the other leg of the triangle be?

A 12 meters

B 15 meters

C 20 meters

D 24 meters

16 Sharon drew a scale drawing of her school building. If the actual building is 52.5 feet tall and Sharon's drawing is 15 inches tall, what scale did Sharon use?

F 1 inch : 52.5 feet

G 1 inch : 4.5 feet

H 1 inch : 3.5 feet

J 2 inches : 15 feet

17 The top of Gabe's desk is 36 inches wide by x inches long. Which expression could Gabe use to find the perimeter of his desk?

A $36 + x$

B $36x$

C $72 + 2x$

D $72x$

18 What is the solution to the equation $15a = 3(3a + 8)$?

F 2

G 3

H 4

J 5

19 Devon paid $29.25 for 3 books. If all the books were the same price, how much would 5 books have cost at the same rate?

A $39.00

B $48.00

C $48.75

D $49.25

20 Which expression is a polynomial with exactly 2 terms?

F $2ab$

G $\frac{2a}{b}$

H $2a + b$

J $a + 2b + ab$

21 Simon has been selling candy to his classmates. On Tuesday, he bought a box of 8 bars of candy for $12. He sold all the candy and made a profit of $20. How much did Simon charge for each candy bar?

A $4

B $3.50

C $3

D $2

22 Simplify the expression $13x + 5y - 9x - 7y$.

F $2xy$

G $18x - 16y$

H $4x + 2y$

J $4x - 2y$

23 What is the fourth term in the following pattern?

1   2   3   ?

A

B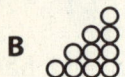

C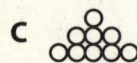

D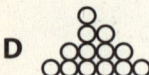

24 The table below shows the number of diagonals that can be drawn from each vertex of an *n*-sided polygon.

Number of Sides, *n*	Number of Diagonals
3	0
4	1
5	2
6	3

Following the pattern shown in the table, how many diagonals could be drawn from each vertex of a polygon with 10 sides?

F 3

G 7

H 8

J 10

PART 2

25 Look at the polygon below:

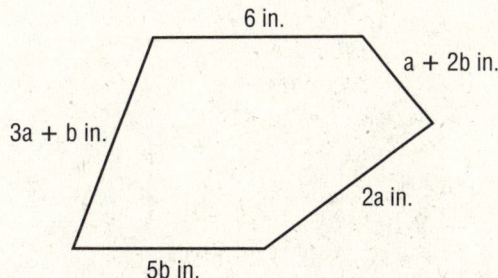

6 in.

a + 2b in.

3a + b in.

2a in.

5b in.

Part A

Write an expression that represents the perimeter of the polygon. Simplify like terms.
Show your work.

Answer _____

Part B

If $a = 3$ and $b = 1.5$, what is the perimeter of the polygon?
Show your work.

Perimeter _____ inches

26 Solve $2(x - 3) = x - 2$

Part A

Solve for x.

Show your work.

Answer _____

Part B

On the lines below, explain why $2(x - 3) = x - 2$ and $x - 3 = \frac{x - 2}{2}$ have the same solution.

_____

_____

_____

27 Look at the pattern of blocks below.

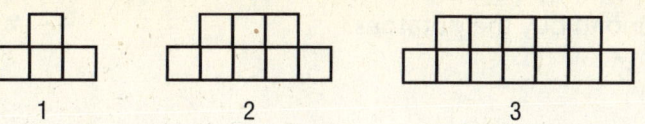

1   2   3   4

Continue the pattern by drawing a graphical representation of the blocks in the fourth pile in the space below.

How many blocks would there be in the fifth pile?

Answer _____ blocks

28 Laura wants to buy 24 potatoes. The supermarket sells bags of 10 potatoes for $2.80, bags of 8 potatoes for $2.40, and single potatoes for $0.35 each. What is the least amount of money Laura can spend to buy the potatoes?

Show your work.

Answer _____

On the lines below, explain the process you used to determine your answer.

_____

_____

_____

_____

_____

29 Explain one method for finding the sum of the interior angles of an octagon.

Part A

Use diagonals to divide the octagon below into triangles.

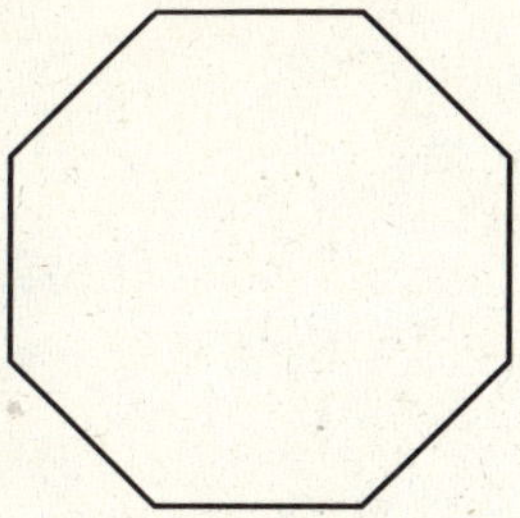

Part B

Using the number of triangles you found as a guide, what is the sum of the interior angles of an octagon?

Answer _____

Part C

Explain your answer on the lines below.

_____

_____

_____

_____

_____

30 Look at the right triangle below.

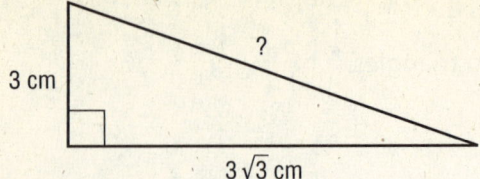

Use the Pythagorean Theorem to find the length of the hypotenuse.
Show your work.

Answer _____ cm

31 Use your ruler to help you solve the problem.

The map uses a scale of 1 inch for every 60 miles. What is the actual distance between Canton and Batavia in miles?

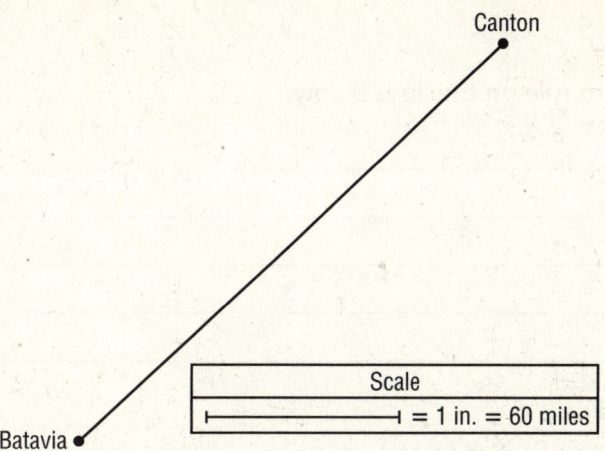

Show your work.

Answer _____ miles

32 Look at the following number pattern:

3	9	27	81	?

Part A

Describe the pattern rule on the lines below.

_____

_____

_____

_____

_____

Part B

What should the fifth term of the pattern be?

Answer _____.

Part C

Now look at this pattern:

2	6	18	54	162

Does it obey the same pattern rule as the first group of numbers? Explain why or why not on the lines below.

_____

_____

_____

_____

_____

33 A dry cleaner uses the following table to calculate how much to charge to clean a particular number of dresses.

Number of Dresses	Cost ($)
1	6
2	10
3	14
4	18
5	22

Part A

What equation describes the relationship between the input and the output in the table above?

Equation _____

Part B

If a woman wants to have 15 of her dresses cleaned, how much should she be charged?
Show your work.

Answer _____

34 A 24-foot ladder is placed against a building so that its foot is 10 feet away from the base of the building. Draw a sketch and determine to the nearest foot how far up the side of the building the ladder reaches. Use a calculator.

Show your work.

Answer _____ feet

35 Each interior angle of a regular polygon measures 108°.

Part A

Write an equation to find the number of sides n in this polygon and solve.

Part B

Let x = the sum of the measures of the interior angles of a pentagon. Write an expression for the sum of the measures of the interior angles of a hexagon using x.

Expression _____

36 Nadia put $65 of her weekly paycheck in a college savings account. Then she bought a T-shirt at the mall for $12 and spent half as much on lunch. She also lent her friend Sara $5 to buy a pair of earrings. After her trip to the mall, Nadia had $18 left. What was the amount of her paycheck?

Show your work.

Part A

Answer _____

Part B

Explain your method.

_____

_____

_____

LESSON 4.1 — Finding and Applying Laws of Exponents

New York Performance Indicators

8.N.1 Develop and apply the laws of exponents for multiplication and division

8.CN.3 Connect and apply a variety of strategies to solve problems

8.PS.4 Observe patterns and formulate generalizations

VOCABULARY

A **base** is a number or variable that is multiplied by itself.

The **exponent** is the number of times the base is used as a factor.

A **power** is a number that can be expressed using an exponent.

REVIEW

Understanding Exponents

You can write $3 \times 3 \times 3 \times 3$ as 3^4.

In the power 3^4, the base is 3 and the exponent is 4. 3^4 is read as *3 raised to the fourth power.*

You can evaluate a power using multiplication.

$$3^4 = 3 \times 3 \times 3 \times 3 = 81$$

Applying Exponents

Exponents are useful to write very large or very small numbers.

Light travels at a speed of 300,000,000 meters every second! Using exponents, you can write the speed of light as follows.

$$3 \times 10^8 \text{ meters per second}$$

Multiplying Powers with the Same Base

When you multiply or divide powers with the same base, you can simplify them using the laws of exponents. The base can be a number or a variable. To find the law or rule for multiplying two powers with the same base, simplify $5^3 \times 5^4$.

$$5^3 \times 5^4 = (5 \times 5 \times 5) \times (5 \times 5 \times 5 \times 5) = 5^7 \text{ or}$$
$$5^3 \times 5^4 = 5^{(3 + 4)} = 5^7$$

This pattern holds true for all nonzero numbers, giving us the following law.

To multiply two powers with the same base, add the exponents.

$$x^a \times x^b = x^{a + b}$$

EXAMPLE 1

Simplify the expression 4×4^4.

You can simplify this problem using the law for multiplying powers with the same base.

The exponent of 4 is 1, because $4^1 = 4$. Since both 4 and 4^4 have 4 as a base, you can simplify the expression by adding the exponents. Then you can evaluate the power by using multiplication.

$$4 \times 4^4 = 4^{(1+4)}$$
$$= 4^5$$
$$= 1,024$$

▶ **Understanding the Solution** The number 1,024 looks pretty big. Are you sure it's right? You can check your result by using the distributive property:

$4^5 = 4^2 \times 4^3 = 16 \times 64 = 16(60 + 4) = 960 + 64 = 1,024$.

TRY IT!

Apply the law for multiplying powers with the same base to simplify the expression $m^4 \bullet m^6$.

Raising a Power to a Power

To find the law for raising a power to a power, simplify $(5^3)^4$. Use 5^3 as a base 4 times.

$$(5^3)^4 = 5^3 \times 5^3 \times 5^3 \times 5^3$$

Now, apply the law of exponents for multiplying powers with the same base.

$$(5^3)^4 = 5^3 \times 5^3 \times 5^3 \times 5^3 = 5^{3+3+3+3} = 5^{12}$$
$$\text{or, } (5^3)^4 = 5^{3 \times 4} = 5^{12}$$

This pattern gives us the following law.

To raise a power to a power, multiply exponents.

$$(x^a)^b = x^{a \cdot b}$$

EXAMPLE 2

Simplify the expression $(s^4)^5$.

You can solve this problem using the law for raising a power to a power.

$$(s^4)^5 = s^{4 \cdot 5} = s^{20}$$

▶ **Understanding the Solution** Since s is a variable, you can only simplify this expression. You cannot evaluate it.

TRY IT!

Apply the law for raising a power to a power to simplify the expression $(2^2)^3$. Evaluate the result.

Dividing Powers with the Same Base

To find the law for dividing two powers with the same base, simplify $\frac{5^3}{5^2}$.

$$\frac{5^3}{5^2} = \frac{5 \times \cancel{5} \times \cancel{5}}{\cancel{5} \times \cancel{5}} = 5$$

Notice that you simplify $\frac{5^3}{5^2}$ by subtracting the exponents: $5^{(3-2)} = 5^1$, or 5. This pattern holds true for all nonzero numbers, giving us the following law.

To divide two powers with the same base, subtract the exponents.

$$\frac{x^a}{x^b} = x^{a-b}, \, x \neq 0$$

EXAMPLE 3

Simplify the expression $\frac{4^8}{4^5}$.

You can solve this problem using the law for dividing powers with the same base.

$$\frac{4^8}{4^5} = 4^{8-5}$$
$$= 4^3$$
$$= 64$$

▶ **Understanding the Solution** You could get the same result by finding the product of eight factors of 4 and dividing that by the product of five factors of 4. However, applying the law for dividing powers with the same base is much quicker.

TRY IT!

Apply the law for dividing powers with the same base to simplify the expression $\frac{x^5}{x^2}$.

Exercises

SHORT RESPONSE

1 On the lines below, explain why $5^4 \times 5^3 = 5 \times 5^6$.

_____

_____

2 Nancy is simplifying the expression $(4^2 \times 4^3)^2$. She thinks that since the bases are the same, she can add all the exponents. On the lines below, explain why you agree or disagree with Nancy's idea. Then simplify the expression.

_____

_____

_____

$(4^2 \times 4^3)^2 = $ _____

3 Which expression equals x^{12}?

A $x^2 \cdot x^6$

B $x^4 \cdot x^3$

C $x^{11} \cdot x$

D $x \cdot x^{12}$

4 Which expression below has a value of 64?

F 4^4

G $\dfrac{4^5}{4^2}$

H $4^2 \times 4^3$

J $(4^2)^3$

5 Which symbol would make the statement true?

$$15^2 \ \square \ (2^5)^3$$

A $<$

B $>$

C $=$

D $\div$

6 What do you do to the exponents to simplify the expression $(b^4)^6$?

F add

G subtract

H divide

J multiply

7 Which exponent would make the equation true?

$$5^6 \times 5^{\square} = 5^{24}$$

A 3

B 4

C 18

D 24

8 What is the value of 10^4 divided by 10^3?

F 10^0

G 10^1

H 10^7

J 10^{12}

9 A box in the shape of a cube has edges that measure y^5 inches. What is the volume of the box in cubic inches?

A y^3

B y^8

C y^{10}

D y^{15}

10 Which power is equal to $3^3 \times 9$?

F 27^3

G 9^4

H 3^5

J 3^4

11 Marcel is doing his math homework. He simplifies the expression $(3^2 \times 4^3)^2$ as follows.

$$(3^2 \times 4^3)^2 = ((3 \times 4)^{(2+3)})^2 = 12^{10}$$

Part A

Did Marcel make a mistake? On the lines below, explain why or why not.

_____

_____

_____

Part B

Marcel now has to simplify $\left(\dfrac{8^3}{2^5}\right)^2$. Explain how 8^3 can be rewritten using the base 2. Then explain how to apply the laws of exponents to simplify Marcel's expression.

_____

_____

_____

Answer _____

LESSON 4.2   Evaluating Expressions with Integral Exponents

New York Performance Indicators

8.N.2   Evaluate expressions with integral exponents

8.PS.3   Understand and demonstrate how written symbols represent mathematical ideas

8.PS.8   Understand how to break a complex problem into simpler parts or use a similar problem type to solve a problem

VOCABULARY

An **integral exponent** represents repeated multiplication.

The **order of operations** tells you which operations to perform first when evaluating expressions. Working from left to right in the expression, use the following order of operations.

1. Evaluate operations in parentheses ( ).
2. Simplify powers.
3. Multiply and divide.
4. Add and subtract.

REVIEW

Understanding Expressions with Exponents

Look at the expression 2×5^2.

The order of operations says you can evaluate this expression by multiplying from left to right. However, you must simplify the power first.

$$2 \times 5^2 \neq 10^2$$

$$2 \times 5^2 = 2 \times 25$$

This makes sense if you write the power out as factors.

$$2 \times 5^2 = 2 \times 5 \times 5$$

What You Should Know

Any nonzero number raised to the 0 power is 1.

$$4^0 = 1$$

Any nonzero number raised to a negative exponent is equal to the reciprocal of the power.

$$4^{-2} = \frac{1}{4^2}$$

If an exponent is an even number, the result will be a positive number if the base is not equal to 0.

$$(-4)^2 = 16$$

If an exponent is an odd number, the result will be a negative number when the base is negative.

$$(-4)^3 = -64$$

Evaluating Expressions with Positive Integral Exponents

When you evaluate or simplify expressions with integral exponents, follow the order of operations.

EXAMPLE 1

Evaluate the expression $(9 + 1)^2 + 2^3 \times 2$.

You can use the order of operations to solve this problem.

Parentheses first	$(9 + 1)^2 + 2^3 \times 2 = 10^2 + 2^3 \times 2$
Exponents next	$= 100 + 8 \times 2$
Multiply/Divide	$= 100 + 16$
Add/Subtract	$= 116$

▶ **Understanding the Solution** Make sure you follow the order of operations. Note that $(9 + 1)^2 \neq 9^2 + 1^2$.

TRY IT!

Evaluate the expression $300 + 10 (1.5 + 3)^2$.

EXAMPLE 2

A number can be expressed as the product of its prime factors. The prime factorization of a number is $2^2 \times 3^2 \times 5$. What is the number?

You can use the order of operations to solve this problem.

$$2^2 \times 3^2 \times 5 = 4 \times 9 \times 5$$
$$= 180$$

▶ **Understanding the Solution** The order of operations is to evaluate each power, then multiply. The expression $2^2 \times 3^2 \times 5$ is equivalent to $2 \times 2 \times 3 \times 3 \times 5$.

TRY IT!

The prime factorization of a number is $3^2 \times 5 \times 7^2$. What is the number?

Evaluating Expressions with Negative Exponents

To simplify an expression with a negative exponent, rewrite the power as a fraction using 1 as the numerator and the power with a positive exponent as the denominator.

$$x^{-n} = \frac{1}{x^n}, x \neq 0$$

EXAMPLE 3

Evaluate the expression 3^{-2}.

You can solve this problem by rewriting the power as a fraction.

$$3^{-2} = \frac{1}{3^2}$$

$$= \frac{1}{9}$$

▶ **Understanding the Solution** $\frac{1}{3^2}$ or 3^{-2} is the reciprocal of 3^2.

EXAMPLE 4

The mass of a bumblebee can be as small as 5×10^{-2} grams. Evaluate the expression 5×10^{-2}.

You can solve this problem by using the rule for simplifying negative exponents.

$$5 \times 10^{-2} = 5 \times \frac{1}{10^2}$$
$$= \frac{5}{100} \text{ or } 0.05 \text{ grams}$$

▶ **Understanding the Solution** You can use negative powers of 10 to write decimal numbers.

TRY IT!

Write 6×10^{-3} as a decimal.

Exercises

SHORT RESPONSE

1 Evaluate the expression 5×3^{-3}.

Show your work.

Answer _____

2 Does the expression $3^2 \times 5^2 - 4^2 + 1$ have the same value as $3^2 \times (5^2 - 4^2) + 1$? Explain why or why not on the lines below.

Show your work.

Answer

_____

_____

_____

3 Evaluate the expression.

$$2 + (2 + 200) \times 2^4$$

A 3,234
B 3,264
C 6,400
D 6,420

4 Which operation should be performed first to evaluate the following expression?

$$(20 + 5)^2 + 4 \times 2^6$$

F simplify 2^6
G add 4
H multiply 4 and 2
J add 20 and 5

5 Which symbol would make the statement true?

$$3 + 15^2 \,\square\, (3 + 15)^2$$

A $<$
B $>$
C $=$
D $\div$

6 Which expression below has a value of 0?

F $(3 \times 1)^4 \times 4^2$
G $(3 \times 0)^4 \times 4^2$
H $(3 + 0)^4 + 4^2$
J $(3 \times 0)^4 + 4^2$

7 Which exponent would make the equation true?

$$4 \times 10^{\square} = 0.004$$

A 3
B 2
C -3
D -2

8 The table below shows the prime factorization of some numbers.

Number	Prime Factors
15	3×5
25	5^2
42	$2 \times 3 \times 7$
b	$2^2 \times 3^2 \times 11$
420	$2^2 \times 3 \times 5 \times 7$

What is the value of the number b?

F 66
G 264
H 396
J 406

9 Distances between planets are measured in astronomical units (AU). One AU is the average distance between the center of the Sun and the center of Earth, which in scientific notation is roughly 9.3×10^7 miles.

Part A

The minimum possible distance between Jupiter and Earth is about 3.92 AU. Write an expression to convert this distance from AU to miles.

Answer _____

Part B

Evaluate the expression you wrote in Part A to find the distance in miles in standard form.

Show your work.

Answer _____ miles

Percents Less than 1% and Greater than 100%

New York Performance Indicators

8.N.3 Read, write, and identify percents less than 1% and greater than 100%

8.N.4 Apply percents to: Tax, percent increase/decrease, simple interest, sale price, commission, interest rates, and gratuities

8.CN.1 Understand and make connections among multiple representations of the same mathematical idea

8.R.8 Use representation as a tool for exploring and understanding mathematical ideas

VOCABULARY

Percent means "per hundred." You can express a part of a whole as a percent. For example, 1 part out of 20 parts is 5%.

A **percent increase** or **percent decrease** is the ratio of the amount of change to the original amount, written as a percent.

REVIEW

Understanding Percents

To find what percent a part is of a whole, divide the whole into 100 equal parts.

A percent can also be written as a fraction or as a decimal.

$$5\% = \frac{5}{100} = 0.05$$

Applying Percents

Suppose 2% of the bulbs made by a manufacturer are defective. How many bulbs will be defective in a shipment of 600 bulbs?

$$2\% \text{ of } 600 = \frac{2}{100} \times 600 = 12$$

12 bulbs will be defective.

Percents Less than 1%

You can visualize percents that are less than 1% of the whole. One out of 100 squares represents 1%. In the figure below, the shaded part represents 0.5% or half of 1%.

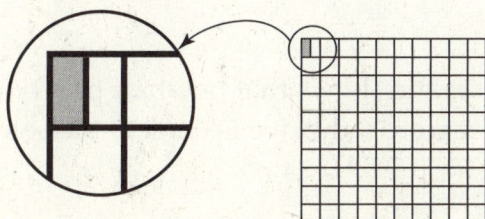

Scientists often work with amounts that are less than 1% of the whole.

EXAMPLE 1

A sample of rock weighing 400 grams contains 3 grams of iron. What percent of the rock sample is iron?

You can also solve a percent problem by using a proportion.

To find the percent, set the ratio $\frac{3}{400}$ equal to the percent ratio, $\frac{x}{100}$.
Now you can solve the proportion for x.

$$\frac{3}{400} = \frac{x}{100}$$

$$400x = 3 \times 100$$

$$x = \frac{3 \times 100}{400}$$

$$x = 0.75$$

▶ **Understanding the Solution** The 3 grams of iron make up 0.75% of the 400-gram rock sample. Notice that this method suggests a shortcut to find percents. Instead of setting up a proportion, you can multiply the original ratio $\frac{3}{400}$ by 100.

TRY IT!

A chemistry teacher sets aside 500 grams of a chemical for a class experiment. If she gives each student 4 grams of the chemical, what percent of the chemical supply will each student have?

EXAMPLE 2

A chemist adds 0.4 mL of dye to a solution. The total volume of the solution is now 80 mL. What percent of the solution is the dye?

You can solve this problem by using equivalent fractions.

Write a ratio comparing the volume of dye to the total volume of the solution. $\frac{0.4}{80}$

Now find an equivalent fraction that expresses this ratio using 100 as a denominator.

$$\frac{0.4}{80} = \frac{4}{800}$$

$$= \frac{1}{200}$$

$$= \frac{\frac{1}{2}}{100}$$

▶ **Understanding the Solution** The last ratio shows that one-half part out of 100 parts (or half of 1%) is the dye. In other words, the dye is 0.5% of the solution. You can also multiply $\frac{0.4}{80} \times 100 = \frac{40}{80} = \frac{1}{2}$, or 0.5%.

TRY IT!

A biologist collects 50 liters of lake water. She sends 0.2 liters of the lake water to a lab to be tested. What percent of the collected lake water does she send for testing?

Percents Greater than 100%

A change in a quantity can be expressed as a *percent increase* or *percent decrease*. If the amount of change is more than the original amount, then the percent increase is greater than 100%.

EXAMPLE 3

In 1950, the world population was 2.6 billion. In 2000, the world population was 6.1 billion. What was the percent increase in population between 1950 and 2000?

You can solve this problem by using the definition of percent increase.

$$\text{percent increase} = \frac{\text{population increase}}{\text{original population}} \times 100$$

$$= \frac{6.1 - 2.6}{2.6} \times 100$$

$$= 135\%$$

▶ **Understanding the Solution** The population increase is greater than 100% because the population more than doubled between 1950 and 2000.

TRY IT!

In 2000, the national census showed that the population of New York State was almost 19 million. The 1900 census showed that the population was about 7.3 million. What was the percent increase in the population of New York between 1900 and 2000?

Exercises

SHORT RESPONSE

1 John put his money into a savings account. He initially deposited $400. After twenty years, he had $900 in the bank. What was the percent increase in his savings?

Show your work.

Answer _____

2 Pure water boils at a temperature of 100°C. A chemist measures the boiling-point temperature of pure water with a thermometer. His first reading is 99.7°C. Find the error in measurement as a percent.

Show your work.

Answer _____

3 A biologist is looking for a rare flower in a park forest. Which ratio represents less than 1% of the whole?

A 1 flower out of 50 flowers
B 2 flowers out of 180 flowers
C 1 flower out of 150 flowers
D 5 flowers out of 500 flowers

4 What is the decimal value of 0.3%?

F 0.003
G 0.03
H 3.0
J 300

5 A cell biologist was growing bacteria in a dish. At Hour 1, there were 200 bacteria in the dish, and at Hour 2 there were 500 bacteria. What percent increase occurred between Hour 1 and Hour 2?

A 50%
B 150%
C 200%
D 250%

6 0.04% is equivalent to which fraction?

F $\frac{4}{10}$

G $\frac{4}{100}$

H $\frac{4}{1,000}$

J $\frac{4}{10,000}$

7 Which symbol makes the relationship true if $x > 0$?

$$127\%x \ \square \ 127x$$

A >
B <
C =
D ≥

8 A population of zebras was grouped by age. The results are shown below:

Age (in years)	Number of Zebras
0–10	65
11–20	430
21–30	170
31–40	35
41–50	5

What percent of the zebras are over 40 years old?

F 0.07%
G 0.7%
H 7%
J 70%

9 The table shows the attendance at a car show. Friday's attendance is more than a 100% increase over the attendance for which day?

Tuesday	2,100
Wednesday	2,065
Thursday	2,177
Friday	4,200

A Tuesday only
B Wednesday only
C Thursday
D none of the days

10 The table shows the membership of three different clubs at a local school. Each student belongs to only one club.

Club	Number of Students
Chess Club	3
Book Club	29
Sport Club	295

Part A

On the lines below, explain how to find the percent of students who belong to the Chess Club. Then write the percent.

_____

_____

Answer _____

Part B

This circle graph compares the size of each club to the total number of students in the school by dividing the circle into sectors. Explain how to find the measure of the angle at the center of the circle that is used for the chess club's sector. Then write the measure of the angle below to the nearest whole degree.

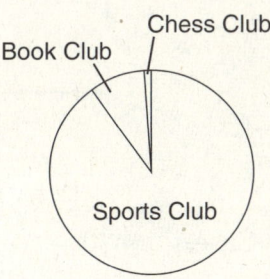

_____

_____

Answer _____

Problem-Solving Strategy:
Testing Conjectures

New York Performance Indicators

8.RP.3 Evaluate conjectures by distinguishing relevant from irrelevant information to reach a conclusion or make appropriate estimates

8.RP.4 Provide supportive arguments for conjectures

8.RP.6 Support an argument by using a systematic approach to test more than one case

A **conjecture** is an educated guess that is based on patterns of results. You must test or evaluate a conjecture to see if it is always true.

Problem: Test the conjecture: *The square of a number is always greater than the number.*

SOLUTION

What do you know?
You can use a number line to see the different possible cases for this conjecture.

What do you need to find?
If the conjecture is true for all numbers.

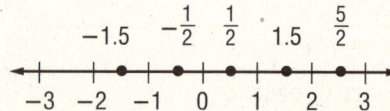

Strategy   Let x represent the number.

Case 1: $x < 0$

$$\left(-\frac{1}{2}\right)^2 = \frac{1}{4} \qquad (-1.5)^2 = 2.25$$

The square of a negative number is always positive. Also, a positive number is always greater than a negative number.

Conclusion: If $x < 0$, then the square of the number is greater than the number. The conjecture is true for negative numbers.

Case 2: $x \geq 0$

Case A: Test numbers from the interval between 0 and 1, or $0 \leq x \leq 1$.

$$0^2 = 0 \qquad \left(\frac{1}{2}\right)^2 = \frac{1}{4} \qquad 1^2 = 1$$

In this interval, the square of the number is either equal to or less than the original number.

Case B: Test positive numbers greater than 1, or $x > 1$.

$$(1.5)^2 = 2.25 \qquad \left(\frac{5}{2}\right)^2 = \frac{25}{4}$$

If $x > 1$, then the square of the number is greater than the number.

Conclusion: The conjecture is true for numbers greater than 1.

▶ **Understanding the Solution**   This conjecture is only true for negative numbers ($x < 0$) and for numbers greater than 1 ($x > 1$). So the conjecture is false.

SHORT RESPONSE

1 Test the conjecture: *A number that is divisible by 10 is also divisible by 2 and 5.* Then write ***true*** or ***false*** on the line below.

Show your work.

Answer ________________

On the lines below, use words to explain why the conjecture is true or false.

____

____

____

____

2 Test the conjecture: *$2^n - 1$ is a prime number for positive integer values of $n > 1$.* Evaluate the conjecture for several possible cases. Then write ***true*** or ***false*** on the line below.

Show your work.

Answer ________________

LESSON 4.5 Applying Percents to Real-World Situations

New York Performance Indicators

8.N.4 Apply percents to: Tax, percent increase/decrease, simple interest, sale price, commission, interest rates, and gratuities

8.CN.4 Model situations mathematically, using representations to draw conclusions and formulate new situations

8.CN.8 Investigate the presence of mathematics in careers and areas of interest

8.R.10 Use mathematics to show and understand social phenomena (e.g., determine profit from sale of yearbooks)

VOCABULARY

The **base** in a percent proportion is the original or whole amount.

The **part** in a percent proportion is the amount being compared to the whole amount.

REVIEW

Understanding Applications of Percents

Percents are often used to calculate and describe tax, simple interest, commission, sale price, and tip.

In all these applications, you can use the following percent equation.

$$\frac{percent}{100} \times base = part$$

You can also use the percent ratio to write a proportion.

$$\frac{percent}{100} = \frac{part}{base}$$

What You Should Know

In many applications, after using the percent equation or ratio, you must add or subtract the amount you got from the base.

Examples

- **Add** the sales tax amount to the price.

- **Add** the interest amount to the original amount.

- **Add** the tip amount to the bill.

- **Subtract** the discount amount from the original price.

Applying Percents to Sales Tax

EXAMPLE 1

Sales tax in New York City is 8.375%. Find the tax you would pay on a video game that costs $39.00, and the total cost of the game including tax.

You can solve this problem by finding the amount of tax and adding it to $39.

To find 8.375% of $39.00, you can use the percent equation with 39 as the base.

$\frac{8.375}{100} \times 39 = \frac{326.625}{100} \approx 3.27$, so the amount of sales tax is $3.27.

Add the sales tax to the price of the video game to get the total cost.

$39.00 + 3.27 = 42.27$, so the game would cost $42.27 including tax.

▶ **Understanding the Solution** You can also find 8.375% of $39 using the percent ratio: $\frac{8.375}{100} = \frac{x}{39}$. Remember, when working with money amounts you should always round the answer to the nearest cent.

TRY IT!

In London, the sales tax is 17.5%. If an order of fish and chips costs £4.00, find the tax and the total price including tax.

Applying Percents to Simple Interest

Simple interest is the amount of money that is paid or earned for using money. **Principal** is the original or base amount of money used to calculate simple interest. To calculate simple interest, you can use the percent equation and multiply by the number of years given in the problem.

EXAMPLE 2

Marco gets a 2-year loan for $5,000 at 11.2% simple interest. Find the interest amount and the total amount he will have to pay on the loan.

You can solve this problem by finding 11.2% of $5,000.

Use the percent equation with 5,000 as the base. $\frac{11.2}{100} \times 5,000 = \frac{56,000}{100} = 560$

This is the amount of interest he would owe in 1 year.

Now multiply your result by 2. $560 \times 2 = 1,120$

Add the total interest amount to the original loan. $5,000 + 1,120 = \$6,120$

▶ **Understanding the Solution** You can also calculate simple interest using the formula $I = prt$, where I is the interest amount, p is the principal (amount deposited or borrowed), r is the interest rate, and t is time in years. In this problem, $p = 5,000$, $r = 11.2\%$, and $t = 2$. So $I = 5,000 \times \frac{11.2}{100} \times 2 = 1,120$, as before.

TRY IT!

Find how much interest Sandra will earn if she puts $2,500 in the bank for 6 years at 2% simple interest. Then find the total amount she will have after 6 years.

EXAMPLE 3

Enrico invests $7,500 in a local business. After 2 years, he has $8,700. How much money did he make, and what was the percent interest (also called the rate of return)?

You can solve this problem using the interest equation, I = prt.

Enrico made $8,700 – $7,500 = $1,200 in interest. This is the interest amount, *I*.

Now substitute 1,200 and the principal, 7,500, into the equation *I = prt* and solve for *r*.

$$1,200 = 7,500 \cdot r \cdot 2$$
$$1,200 = 15,000r$$
$$\frac{1,200}{15,000} = r$$
$$r = 0.08, \text{ so he had a return of } 8\%.$$

▶ **Understanding the Solution** The interest earned in 1 year is $1,200 ÷ 2 = $600. To find what percent 600 is of 7,500, you could set up the proportion $\frac{p}{100} = \frac{600}{7,500}$ and solve for *p*, the unknown percent. $p = \frac{600}{7,500} \times 100 = 8\%$

TRY IT!

Barbara borrows $20,000 from the bank to buy a car. At the end of 6 years, she has paid a total of $32,000. How much interest did she pay, and what was the simple interest rate?

EXAMPLE 4

A realtor earns 6% commission on a $350,000 house she sells. How much money does she make?

You can solve this problem by finding 6% of 350,000.

$\frac{6}{100} \times 350,000 = \frac{2,100,000}{100} = 21,000.$ She makes $21,000.

▶ **Understanding the Solution** Most realtors make money only if they sell a house. In some jobs, however, a commission or percent of sales is added onto a base pay or salary.

TRY IT!

A clerk at an electronics store makes $200 per week, plus 4% of any sales. If he sells $3,000 worth of electronic items one week, how much will he get paid?

Exercises

SHORT RESPONSE

1 A scarf is originally priced at $30.00. It is then discounted by 25%. Several days later, it is discounted 10% off the new price. Is the cost of the scarf the same as if the store had taken 35% off the original $30? Explain your answer on the lines below.

_____

_____

_____

2 Ann and Becky spend $70.00 on dinner. If they want to leave a 15% tip, find the amount of tip they should leave.

A $0.15
B $1.50
C $10.50
D $80.50

3 Yuri makes $8.00 an hour, plus a 3% commission on sales. If he works for 30 hours and sells $4,600 worth of merchandise, how much money will he earn?

F $378.00
G $1,389.60
H $1,700.00
J $4,632.00

4 A realtor who gets a 5% commission makes $7,500 on the sale of a house. What was the price of the house?

A $35,000
B $70,000
C $140,000
D $150,000

5 There are 340 students at Walker Elementary this year. Next year, the school is predicting a 20% increase in students. If the prediction is correct, how many students will the school have next year?

F 68
G 360
H 408
J 680

6 A hairbrush sells for $4.25 and tax is 8.375%. What is the total cost?

A $3.56
B $4.29
C $4.33
D $4.61

7 Nick buys a car for $15,000. He gives the car dealer $5,000 and borrows the rest of the money at 7% simple interest for 5 years. What is the total amount of interest he will have to pay on the loan?

F $350
G $1,750
H $3,500
J $5,250

8 Mika buys a dress for $43 that was marked down from $68. What is the percent discount off the original price?

A 25%
B 37%
C 58%
D 63%

9 A $400,000 mortgage must be paid over 25 years. If the simple interest paid is $540,000, which expression could you use to find the interest rate on the mortgage?

F $r = \dfrac{540,000}{(400,000)(25)}$

G $r = \dfrac{540,000}{(400,000 + 31,600)} \cdot 25$

H $r = \dfrac{(540,000)(25)}{400,000}$

J $r = \dfrac{(400,000)(25)}{540,000} \cdot 100$

10 Compound interest is interest applied to the sum of the principal amount and any interest already earned. If an account is compounded annually (once every year), the accumulated interest from the previous year is added to the principal amount, and the account earns interest on the "new" principal.

Part A

Rahul invests $5,000 in an account with a fixed 2% interest rate, compounded annually. How much will Rahul have in 3 years if he does not invest more money in the account?

Show your work.

Answer _____

Part B

Suppose Rahul invests $5,000 at a 2% simple interest rate for 3 years. Will the money earn more or less interest than it would at a 2% compound interest rate (compounded annually)?

Show your work.

Answer _____

LESSON 4.6 Estimating Percents

New York Performance Indicators
8.N.5 Estimate a percent of a quantity, given an application
8.PS.2 Construct appropriate extensions to problem situations
8.PS.12 Interpret solutions within the given constraints of a problem
8.RP.3 Evaluate conjectures by distinguishing relevant from irrelevant information to reach a conclusion or make appropriate estimates

VOCABULARY

Estimation is a way to find an approximate answer instead of an exact one.

REVIEW

Understanding Common Percents

You can find many common percents mentally, without using a calculator. Look at the equivalent fractions or numbers for the common percents listed below.

$10\% = \frac{10}{100} = \frac{1}{10}$ $100\% = \frac{100}{100} = 1$

$25\% = \frac{25}{100} = \frac{1}{4}$ $200\% = \frac{200}{100} = 2$

$50\% = \frac{50}{100} = \frac{1}{2}$

What You Should Know

- To find 10% of a number, move the decimal point one place to the left in the number.

- To find 25% of a number, divide the number by 4.

- To find 50% of a number, divide the number by 2.

- 100% of a number is the number itself.

- To find 200% of a number, double the number.

Using Common Percents to Estimate Percents

EXAMPLE 1

Sales tax in New York City is 8.375%. Estimate the total price, including tax, for a shirt with a sticker price of $42.00.

You can solve this problem by finding 10% of $42.00, since 10% is close to 8.375%.

To find 10% of $42.00, move the decimal in 42.00 one place to the left. The tax will be approximately $4.20.

To get the approximate total price, add $42.00 + $4.20 = $46.20.

▶ **Understanding the Solution** Finding 10% of 42.00 is the same as multiplying $\frac{1}{10} \times 42$ or dividing 42 by 10. Remember, moving the decimal point is a quick way to divide or multiply by a power of 10. The exact tax would be $\frac{8.375}{100} \times 42 \approx 3.52$, or $3.52.

TRY IT!

Gregor takes out a car loan from the bank for $12,500 at 11.2% simple interest. What common percent could you use to estimate the amount of interest he would pay in a year? Find the approximate amount of interest.

EXAMPLE 2

A pair of shoes with a sticker price of $47.00 is marked 25% off. About how much will the shoes cost?

You can solve this problem by finding a compatible number close to 47 and dividing by 4.

Both 44 and 48 are divisible by 4.

$44 \div 4 = 11$ and $48 \div 4 = 12$, so the amount of the discount will be between $11 and $12.

Subtract to find the discounted prices.

$47 - 11 = 36$, and $47 - 12 = 35$, so the sneakers will cost between $35 and $36.

▶ **Understanding the Solution** Finding 25% or one-fourth of 47.00 is the same as multiplying $\frac{1}{4} \times 47 = \frac{47}{4} = \$11.75$. The exact answer is $\$47.00 - \$11.75 = \$35.25$, which falls between the two estimated amounts.

TRY IT!

Use compatible numbers to estimate how much you would have to pay for a CD player that was marked 25% off a price of $33.

Using Common Percents to Find Additional Percents

If you know how to find 10%, 25%, and other common percents without a calculator, you can easily calculate additional percents based on these amounts.

EXAMPLE 3

Suppose you go out to lunch and the check for your meal comes to $15.30. You want to leave a generous 20% tip. How much should you leave for a tip?

You can solve this problem by thinking of 20% as 2 times 10%.

First, round $15.30 down to $15.00.

To find 10% of 15, move the decimal one place to the left to get 1.50.

Now double that amount to get 20%.

$1.50 \times 2 = 3$, so you should leave a tip of $3.00.

▶ **Understanding the Solution** 20% is the same as $\frac{1}{5}$, so you could also estimate the tip by dividing $15 by 5: $15 \div 5 = 3$, or $3.00.

TRY IT!

A hotel brochure claims that it has had a 30% increase in customers in the past year. If they had 2,560 customers last year, use common percents to find exactly how many customers they had this year.

EXAMPLE 4

A drug store purchases a bottle of vitamins for $3.40 wholesale and then adds a 150% markup before they sell it. Use common percents to find the cost of the vitamins after the markup has been added in.

You can solve this problem by finding 50% of $3.40 and 100% of $3.40.

To find 50% of 3.40, divide by 2: $3.40 \div 2 = 1.70$. 100% of 3.40 is 3.40.

Add the two amounts. $1.70 + 3.40 = 5.10$, so a 150% markup of $3.40 is $5.10.

Now add the markup amount to the original wholesale cost. The vitamins will be sold in the store for $3.40 + $5.10 = $8.50.

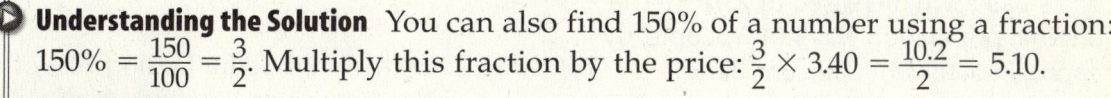 **Understanding the Solution** You can also find 150% of a number using a fraction: $150\% = \frac{150}{100} = \frac{3}{2}$. Multiply this fraction by the price: $\frac{3}{2} \times 3.40 = \frac{10.2}{2} = 5.10$.

TRY IT!

A baseball player had a 210% pay raise added to his salary. Use common percents to find his new salary if his original salary was $200,000. Include the percents you used in your answer.

Exercises

SHORT RESPONSE

1 Jose makes $15,000 a year working part-time. One year, Jose gets a 10% pay cut taken from his salary. The next year, he gets a 10% raise, calculated using this new salary. Is he back to the original salary? Write and explain your answer on the lines below.

_____

_____

2 Car accidents at a busy intersection have increased by 74% in one year. About how many more accidents have occurred, if there were 250 accidents last year? Round your answer up to the nearest whole number.

Show your work.

Answer _____ accidents

3 A realtor gets a 6% commission on the sale of a house. Estimate the amount of commission he earns for selling a $152,000 house.

A $800
B $8,000
C $15,200
D $80,000

4 A newspaper reports that the population of a small town increased to 2,018. If the original population was 1,000 people, by approximately what percent did the population increase?

F 100%
G 125%
H 150%
J 200%

5 The cost of a pair of sneakers has increased 125% from 25 years ago, when they sold for $60. Which method could you use to find the increase in cost?

A Divide 60 by 4.
B Divide 60 by 4 and add the result to 60.
C Multiply 60 by 0.125.
D Multiply 60 by 12.5.

6 Rita got $80 off the price of a new TV. If the $80 was a 25% discount, find the original cost of the TV.

F $20
G $100
H $105
J $320

7 To estimate 23% of a number without a calculator, what fraction should you multiply the number by?

A $\frac{1}{5}$

B $\frac{1}{4}$

C $\frac{1}{2}$

D $\frac{2}{1}$

8 Based on the method for finding 10% of a number, how could you find 1% of a number?

F Move the decimal 1 place to the right.
G Move the decimal 2 places to the right.
H Move the decimal 1 place to the left.
J Move the decimal 2 places to the left.

9 50% of a number is 30. To find the missing number, what could you multiply 30 by?

A $\frac{1}{2}$

B $\frac{2}{1}$

C 0.50

D 50

10 Damen and Nadia want to find 33% of $1,900.

Part A

Damen estimates the percentage by dividing $1,900 by 3. Explain why this method results in a good estimate.

_____

_____

Part B

Nadia thinks that she can find a better estimate by rounding 33% down to 30%, and rounding $1,900 up to $2,000. Whose estimate is better, Damen's or Nadia's?

Show your work.

Answer _____

LESSON 4.7 Using Estimations to Verify Solutions

 New York Performance Indicators

8.N.6 Justify the reasonableness of answers using estimation

8.RP.7 Devise ways to verify results or use counterexamples to refute incorrect statements

8.CM.1 Provide a correct, complete, coherent, and clear rationale for thought process used in problem solving

---

REVIEW

Understanding Estimates

You can estimate by rounding or by finding compatible numbers. A good estimated answer is very close to the exact answer.

Rounding

exact numbers: $2.6 \times 4.25 = $ **11.05**
rounded numbers: $3 \times 4 = $ **12**

Finding compatible numbers

exact numbers: $19 \div 4.25 \approx $ **4.47**
compatible numbers: $20 \div 5 = $ **4** or
$20 \div 4 = $ **5**

Applying Estimates

Cathy wants to buy a shirt for $12.50, pants for $32.99, and socks for $3.75. She estimates that $60 should cover the cost of the clothing before taxes. Is Cathy correct?

exact numbers:
$12.50 + 32.99 + 3.75 < 60$

rounded numbers:
$13 + 33 + 4 = 50$

$\$50 < \$60$, so Cathy is correct.

Estimation by Rounding or Compatible Numbers

When you estimate, you use numbers that are easier to work with than the actual numbers in a problem. You can often use estimation to check an exact answer.

EXAMPLE 1

Luis bought 4 oranges at $0.79 each and a roll of paper towels for $1.29. He paid with a $20 bill and got $14.55 back. Use estimation to explain why the change he got back cannot be correct.

You can estimate the correct change by rounding each number.

The price of oranges can be rounded to $0.80 and the paper towels to $1.30.

$$(4 \times 0.80) + 1.30 = 3.20 + 1.30 = \$4.50$$

The approximate change from $20 is $20 - \$4.50 = \$15.50$, so the change Luis got was about a dollar off.

▶ **Understanding the Solution** Using $0.80 instead of $0.79 makes it much easier to multiply by 4. The addition and subtraction are also much easier with rounded numbers.

Miko buys 6 cans of juice for $0.87 each and a bag of pretzels for $1.49. If she pays with a $10 bill, use estimation to check whether $3.83 is the correct amount of change.

EXAMPLE 2

Two students are working on the same division problem, 542 ÷ 17.5. Rachel gets an answer of 30.97 and Ann gets an answer of 309.7. Use estimation to find out who is probably correct.

You can solve this problem by using numbers that are easier to divide.

Round the number 542 to 500 and round 17.5 to 20. Now you can do the division problem 500 ÷ 20 = 25. This is closer to Rachel's answer, so Rachel is probably correct.

▶ **Understanding the Solution** You could round 17.5 to 18 and 542 to 540 to get a more accurate estimate. However, since Rachel and Ann's answers are so different, you can use the most compatible numbers with more zeros to estimate.

TRY IT!

Two students are trying to find the square root of 130. Jemael thinks it should be about 9.5, and David thinks it is closer to 11.5. Use estimation to find out who is probably correct.

Estimation by Picturing Parts of an Item

Crowd size is often estimated by counting the number of people in part of the crowd, then multiplying by how many parts you see.

EXAMPLE 3

The picture shows the rough size of a crowd at a concert in Central Park. There are about 12 people in each of the squares. The Parks Department estimates that there are 400 people in the crowd. The concert organizers estimate 900 people, and the police department estimates 600. Which estimate is most accurate? Justify your answer.

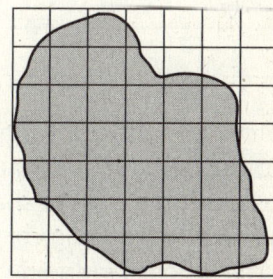

You can solve this problem by counting about how many squares the crowd covers.

Approximately 25 whole squares are shaded. Now combine partially shaded squares to make whole squares. **Keep track of what you are doing by crossing out squares you have counted.** You should end up with a total of about 33 to 36 squares.

$$33 \times 12 = 396 \qquad\qquad 36 \times 12 = 432$$

The crowd size is somewhere between 396 and 432. The Parks Department's estimate was the most accurate.

▶ **Understanding the Solution** It is not usually possible to find the exact number of people in a crowd unless you count them all. It is more practical to get an estimate.

TRY IT!

Suppose the rectangle at the right is a cookie sheet. If approximately 6 cookies fit on the shaded part, how many cookies fit on the sheet?

Exercises

SHORT RESPONSE

1 The table to the right shows the populations of five counties in New York State in 2003. A newspaper claims that in 2003, the total number of people in these five counties was over 2,200,000.
Use estimation to decide if the newspaper is correct. Write and explain your answer on the lines below.

Show your work.

County	Population
Albany	297,845
Allegany County	50,562
Bronx County	1,363,198
Broome County	199,360
Cattaraugus County	83,354

2 The picture to the right shows the plans for a sandbox being made for a park. It takes 5 bags of sand to fill the shaded part of the picture. Will 50 bags of sand be enough to fill the whole sandbox?

Write and explain your answer on the lines below.

3 The total area of the pizza shown below is about 95 square inches.

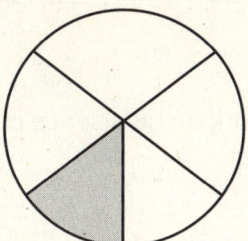

What is the best estimate for the area of the shaded slice?

A 12 square inches

B 19 square inches

C 24 square inches

D 30 square inches

4 For the first week of school, Gabrielle bought 8 boxes of pens at $2.29 each, 4 notebooks at $2.99 each, and 7 boxes of pencils at $5.65 each. Which is the best estimate for her total purchase?

F $63

G $70

H $80

J $88

5 There are 8 red jelly beans in a small 6-ounce bag. Approximately how many red jelly beans will there be in a 32-ounce bag?

A 0

B 24

C 40

D 48

6 Pennies have been scattered evenly over the grid below.

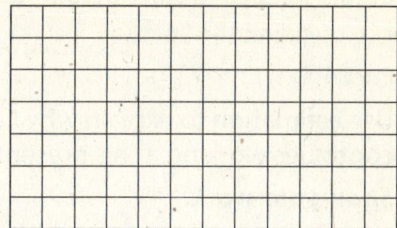

If 4 squares in the grid contain about 51 pennies, approximately how many pennies cover the entire grid?

F 100

G 200

H 500

J 1,000

7 Use estimation to find the approximate answer.

$$6.02 \times 112.7$$

A 6

B 66

C 660

D 6,600

8 A car gets 22.6 miles per gallon of gas. About how many gallons will be needed for a 6,320-mile trip?

F 3

G 31

H 310

J 12,000

9 Luis has 28 math problems to do for his homework. The first problem took 11 minutes and 43 seconds to complete. If he finishes his homework by 8 P.M., he can spend 2 hours on his science project before he has to go to bed at 10 P.M. It is now 3:57 P.M.

Part A

Use estimation to explain why Luis will **not** be able to work on his science project if he continues working at his present rate.

Show your work.

Write your explanation on the lines below.

_____

_____

Part B

Estimate the number of minutes Luis can spend on each problem if he is going to finish his math homework by 8 P.M.

Show your work.

Answer _____ minutes

Explain your answer on the lines below.

_____

_____

_____

Problem-Solving Strategy:
Applying Different Strategies to a Problem

New York Performance Indicators

8.PS.7 Understand that there is no one right way to solve mathematical problems but that different methods have advantages and disadvantages

Sometimes you can solve a problem by using different strategies.

Example: Allen sold 25 old CDs at CD Exchange for $165. The clerk paid him with $5, $10, and $20 bills. If Allen received a total of 10 bills, how many of each bill did he get?

SOLUTION

What do you know?
Allen received $165 in a combination of $20, $10, and $5 bills.

The total number of bills was 10.

What do you need to find?
The number of bills of each denomination

First Method: Make a good guess and check your result.

Try five $20 bills, four $10 bills, and one $5 bill. $\longrightarrow$ $5 \times \$20 + 4 \times \$10 + 1 \times \$5 = \$145$
This is less than $165, so adjust your guess up.

Try seven $20 bills, two $10 bills, and one $5 bill. $\longrightarrow$ $7 \times \$20 + 2 \times \$10 + 1 \times \$5 = \$165$
This is right.

Second Method: You can also solve this problem by writing an equation and looking for its solution.

$$5n + 10m + 20p = 165$$

n is the number of $5 bills, m is the number of $10 bills, and p is the number of $20 bills. Remember that the total number of bills is 10. So if $n = 1$ and $m = 3$, then p has to be 6 because $1 + 3 + 6 = 10$.

From the equation, you can tell that there is at least one $5 bill because the sum is $165. Most of the bills should be $20 because Allen didn't receive very many bills. Try $n = 1$ and $p = 7$, which means $m = 2$.

$$5(1) + 10(2) + 20(7) = 5 + 20 + 140 = 165$$

▶ **Understanding the Solution**   Six $20 bills, four $10 bills, and one $5 bill also add up to $165. However, this is not a solution because the combination uses 11 bills. Remember that the total number of bills should be 10.

1 The product of two consecutive even numbers is 440. What are the numbers?

   Show your work.

   Answer _____

2 Karen has $20 to buy snacks for a party. A bag of chips costs $1.50, a box of cookies costs $2.25, and a carton of juice is $3.75. She bought twice as many bags of chips as boxes of cookies. She got $2 back in change. How many of each item did Karen buy? Make a table to organize your work. Write your answers on the lines provided below.

   Show your work.

   bags of chips: _____

   boxes of cookies _____

   cartons of juice: _____

MULTIPLE CHOICE

1 The eighth graders at Riverside School were grouped according to how many siblings they have. The results are shown below:

Number of Siblings	Number of Students
0	116
1	159
2–3	73
4–6	18
7+	3

What percent of the eighth graders have more than 6 siblings?

A 0.008%

B 0.08%

C 0.8%

D 8%

2 Which exponent would make the equation true?

$$14 + 9^{\square} = 14\frac{1}{729}$$

F −3

G −2

H 2

J 3

3 A galaxy has about two billion stars. If 1% will turn into black holes, approximately what number of stars will turn into black holes?

A 200,000,000

B 20,000,000

C 2,000,000

D 200,000

4 Yoo-Mi buys 4 plums at $0.69 each, 3 bananas at $0.99 each, and a package of paper plates for $2.19. What is the best estimate of the total amount of money she spends?

F $5.80

G $7.00

H $8.00

J $9.20

5 Simplify the expression $\frac{(8^2 \times 8^7)^3}{8^{15}}$.

A 8^{-3}

B 8^2

C 8^{12}

D 8^{27}

6 Jeans were on sale for 35% off the original price. Luis bought a pair of jeans on sale that had an original price of $40.00. How much did Luis pay for his jeans?

F $36.00

G $30.00

H $26.00

J $14.00

7 Evaluate the expressions $3^3 + 2^4 \times 5^2 - 5$ and $3^3 + (2^4 \times 5^2) - 5$.

Show your work.

$3^3 + 2^4 \times 5^2 - 5 = $ _____ $3^3 + (2^4 \times 5^2) - 5 = $ _____

Do the expressions have the same value? Explain why or why not.

_____

_____

_____

8 Rob eats dinner at the grand opening of Le Maison Francais. To celebrate the opening, the restaurant takes 40% off the regular price of every meal. The regular price of Rob's food is $32.00. A sales tax of 6.25% is applied after the restaurant's discount, and Rob leaves an estimated 20% tip on the cost of his meal plus tax. What is the total amount of money Rob spends on his meal?

Show your work.

*Total cost of meal $*_____

Writing and Solving Inequalities

New York Performance Indicators

8.A.1 Translate verbal sentences into algebraic inequalities

8.PS.3 Understand and demonstrate how written symbols represent mathematical ideas

8.PS.6 Represent problem situations verbally, numerically, algebraically, and graphically

---

VOCABULARY

An **inequality** is a mathematical sentence that uses the symbols $<, >, \leq,$ or $\geq$ to compare two amounts or expressions.

The **solution set of an inequality** is the set of values that make the inequality true.

REVIEW

Understanding Inequalities

Inequalities can be graphed on a number line.

$x < 2$, or all numbers less than 2:

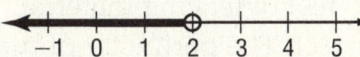

The open circle at 2 means that the 2 is not on the graph.

$x \geq 3$, or all numbers greater than or equal to 3:

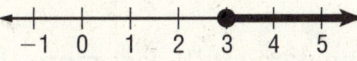

The solid circle at 3 means that the 3 is on the graph.

Applying Inequalities

Paulina charges $10 an hour to babysit. Sometimes she gets tips in addition to her hourly pay. If she made $80 over the weekend, what is the greatest number of hours (*h*) that she could have spent babysitting?

$$h \leq \frac{80}{10}$$
$$h \leq 8$$

Paulina babysat for 8 hours or less.

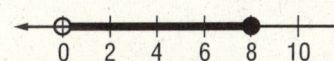

The graph shows that negative values of *h* do not apply in this situation.

Writing Inequalities

You can write inequalities to model real-world situations. To describe a relationship as an inequality, write expressions for the amounts you are comparing. Then look for the following key words in the sentence or problem. Place the matching symbol between the expressions you wrote.

	greater than more than bigger than	less than fewer than smaller than	greater than or equal to larger than or equal to at least not less than	less than or equal to smaller than or equal to at most not more than
Key words				
Symbol	>	<	≥	≤

EXAMPLE 1

Matt has $36 in his savings account. He plans to deposit $5 into his savings account every week. When will he have at least $110 in his savings account? Write an inequality that shows when Matt will have at least $110.

You can write the inequality using expressions for the amounts you are comparing.

First represent the unknown quantity, the number of weeks, by a variable, say w.

Write the expression for the money Matt will have in w weeks.

$$36 + 5w$$

Matt wants to know when he will have *at least* $110. The key word *at least* corresponds to $\geq$.

$$36 + 5w \text{ is } at\ least\ \$110$$
$$36 + 5w \geq \$110$$

▶ **Understanding the Solution** Notice that the order of the amounts you are comparing is important. $36 + 5w \geq \$110$ is not the same as $110 \geq 36 + 5w$.

TRY IT!

Sue wants to fence her rectangular garden. She has 85 feet of fencing material. The length of the garden is 24 feet and the width is w feet. The perimeter of the garden is at most the length of the fencing material Sue has. Write an inequality that compares the perimeter of the garden and the length of the fencing Sue has.

EXAMPLE 2

A new computer game will go on sale next week. The game will cost more than $15, but not more than $23. Write an inequality to express the price range p that the store might charge for the game.

You can write an inequality using more than one inequality symbol.

You are comparing three amounts: the lowest possible price, the actual price, and the highest possible price of the game. The key words in the problem are <u>more than</u> ($>$) and <u>not more than</u> ($\leq$).

To write one inequality for both of these situations, order the amounts you are comparing from the least amount to the greatest amount. Then add the correct symbols.

$$\$15 < p \leq \$23$$

▶ **Understanding the Solution** You can also use two separate inequalities joined by the word "and": $p > \$15$ **and** $p \leq \$23$.

Mr. Pendle is making photocopies for his class. He knows that he has made at least 34 copies, and less than 56 copies. Write an inequality to represent the number of copies that Mr. Pendle may have made.

Solving and Graphing Inequalities

Solving an inequality is similar to solving an equation. In both cases, a variable is isolated on one side of the inequality or equal sign to determine its value. You can then graph the solution of the inequality on a number line.

EXAMPLE 3

Solve the inequality $4x + 2 \geq 18$ for x and graph the solution.

You can solve the inequality the same way you solve equations.

Solve the same way you would solve the equation $4x + 2 = 18$. Perform the same operations on both sides of the inequality sign to isolate the variable.

$$4x + 2 - 2 \geq 18 - 2$$
$$4x \geq 16$$
$$\left(\frac{1}{4}\right)4x \geq 16\left(\frac{1}{4}\right)$$
$$x \geq 4$$

Since x is greater than or equal to 4, the graph of the solution should include $x = 4$. On the number line, this is shown by a solid circle at 4.

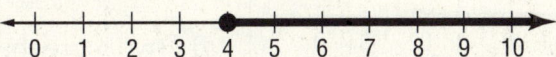

▶ **Understanding the Solution** To check your solution, pick one value within its range and substitute it into the original inequality. Try $x = 5$: $4(5) + 2 \geq 18$ simplifies to $22 \geq 18$, which is true. Now try $x = 3$: $4(3) + 2 \geq 18$ simplifies to $14 \geq 18$, which is not true.

TRY IT!

Solve the inequality $9k - 4 < 23$ for k and graph the solution.

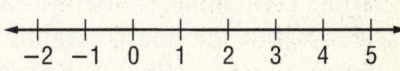

Exercises

1 If $x > \frac{1}{4}$, which inequality symbol belongs in the box? $-15x - 12 \ \square \ 9x - 18$

Show your work.

Answer _____

2 What is the solution of the inequality below?

$$2g + 2 \le -4$$

A $g \le -1$
B $g \ge -1$
C $g \le -3$
D $g \ge -3$

3 Which of the following inequalities correctly expresses the values shown on the number line below?

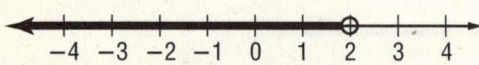

F $z > 2$
G $z < 2$
H $z \ge 2$
J $z \le 2$

4 A car rental company charges \$25/day and 5 cents/mile. Laura rents a car for 2 days, but she doesn't want to spend more than \$65. What is the greatest number of miles she can drive?

A 15
B 30
C 150
D 300

5 Which number line below is the graph of $y \ge -4$ and $y < 6$?

F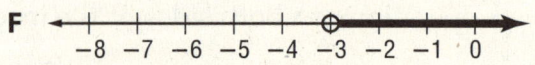

G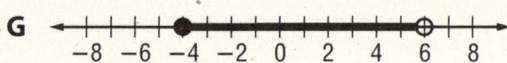

H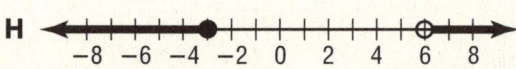

J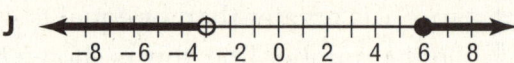

6 Two more than three times a number is at most 37. Which inequality can be used to find the number?

A $37b < 15$
B $3n + 2 \le 37$
C $3b \le 37 + 5b$
D $2 + 3n > 37$

7 Jana is designing a rectanglar garden for her backyard. The length of the garden is 18 feet. She wants the length to be at least 3 feet more than twice the width. Which inequality represents the width of the garden?

F $18 < 2w + 3$
G $3 + 2w > 18$
H $18 \ge 3 + 2w$
J $18 \le 3 - 2w$

8 Suppose 10 less than twice a number is at least 42.

Part A

Use the variable *x* to write an inequality that describes the situation.

Answer _____

Part B

Solve the inequality for *x*.

Show your work.

Answer _____

Part C

Now graph the inequality in the space below.

LESSON 5.2 Communicating Mathematical Expressions in Words

New York Performance Indicators

8.A.2 Write verbal expressions that match given mathematical expressions

8.PS.6 Represent problem situations verbally, numerically, algebraically, and graphically

8.CM.10 Use appropriate language, representations, and terminology when describing objects, relationships, mathematical solutions, and rationale

REVIEW

Understanding Forms of Expressions

An expression can be written with numbers and symbols, or in words. Expressions that match use the same operations and numbers to describe the same amount.

$4 + $3 is the SAME as "three dollars more than four dollars" or "four dollars plus three dollars."

Applying Expressions

Leondra read p pages yesterday. The number of pages she will read today is $4p - 3$. Write this amount in verbal form.

She will read four times the number of pages she read yesterday, minus 3 pages.

OR

The number of pages she will read is 3 less than 4 times the number she read yesterday.

Writing Mathematical Expressions in Words

You can write or read any mathematical expression using words. First, separate the terms of the expression and write the words that describe each term. Then put the verbal forms of the terms together in a sensible order to make a verbal sentence.

Look at how some operations are described below.

x "a number" *or* "a number x"

$\frac{x}{3}$ "a number divided by 3"

$\frac{1}{2}x$ "half of a number"

$2x$ "two times a number" *or* "twice a number"

x^3 "a number cubed" *or* "a number raised to the third power"

$x - 1$ "one less than a number"

EXAMPLE 1

Araf had some bottle caps b last week. He then sold some and bought some new ones. He now has $7b + 2$ bottle caps. Use words to express the number of bottle caps he has now.

You can use the given information to state what the variable b represents.

Start by separating the terms and writing a verbal expression for each.

7b: "seven times the number of bottle caps he had last week"

+ 2: "two more than"

Now make a verbal sentence: Araf has <u>two more than seven times the number of bottle caps he had last week.</u>

▶ **Understanding the Solution** The phrase "two more than" belongs <u>before</u> the phrase that describes 7b because the reverse order does not make a meaningful sentence.

TRY IT!

Leslie has g board games. Tomorrow, she will buy some new games. Then she will have $2g + 5$ games. Express the number of games she will have tomorrow in verbal form.

EXAMPLE 2

Wesley ate c carrots yesterday. Today, he ate $5c - 2$ carrots. The total number of carrots that Wesley ate yesterday and today is equal to T. Write an algebraic expression for T and explain the value of T in words.

You can write an algebraic expression for T and then translate it into words.

Write an expression for T using numbers and symbols. $T = c + (5c - 2)$

Simplify the expression. $T = 6c - 2$

To explain the value of T in words, change the written symbols to verbal expressions.

6c: "six times the number of carrots he ate yesterday"

− 2: "two less than"

T is the total number of carrots that Wesley ate yesterday and today. <u>The total number of carrots is equal to two less than six times the number of carrots he ate yesterday.</u>

▶ **Understanding the Solution** Since the number of carrots Wesley ate yesterday and today are expressed using the same variable c, both the algebraic and verbal expressions for the value of the total can be simplified.

TRY IT!

Jodie toasted m marshmallows last night and $4m + 7$ marshmallows tonight. If S is the total number of marshmallows she toasted, write an algebraic expression for the value of S and explain its value in words.

Explaining Solutions to Problems

EXAMPLE 3

Kaitlin has c charms on her bracelet. Maria has $3c - 5$ charms on her bracelet. If they have a total of 43 charms, how many charms does Maria have? Express your answer both numerically and in verbal form.

You can solve this problem by writing and solving an equation to find c.

Write an equation using the information given in the problem and solve for c, the number of charms Kaitlin has.

$$c + (3c - 5) = 43$$
$$4c - 5 = 43$$
$$4c = 48$$
$$c = 12$$

To find how many charms Maria has, substitute 12 for c in the given expression.

$$3c - 5 = 3(12) - 5 = 36 - 5 = 31$$

Maria has 31 charms, which is five less than three times the number of charms that Kaitlin has.

▶ **Understanding the Solution** To express your solution in verbal form, you must translate the expression $3c - 5$. By describing your solution in terms of c, you explain where the 31 came from.

TRY IT!

Brett sold c boxes of cookies, and Amy sold $2c - 14$ boxes of cookies. If together they sold 37 boxes, how many boxes of cookies did Amy sell? Express your answer both numerically and in verbal form.

Exercises

SHORT RESPONSE

1 Let j represent the number of jelly doughnuts that Randy made this morning. He made $3j - 7$ jelly doughnuts this afternoon. If $j = 23$, how many jelly doughnuts did he make this afternoon?

Show your work.

_____ jelly doughnuts

Express your answer in words using j on the lines below.

_____

_____

2 Andy used *w* wipes to clean the chalkboard on Wednesday. On Friday, it took him $9w - 3$ wipes. Which verbal expression describes the number of wipes Andy had to use to clean the board on Friday?

A three more than nine times the number of wipes he used on Wednesday

B nine times more than five less than the number of wipes he used on Wednesday

C three less than nine times the number of wipes he used on Wednesday

D nine less than three times the number of wipes he used on Wednesday

3 Which expression can be read as "forty-four more than negative twenty-one times a number *n*"?

F $-21 - 44n$

G $44n + 21$

H $21 - 44n$

J $-21n + 44$

4 On Saturday, Danny did *s* sit-ups. On Sunday, he did fewer than eighteen less than nine times the number of sit-ups he did on Saturday. If the number of sit-ups he did on Friday and Saturday is equal to 55, and he did 32 sit-ups on Friday, which of the following could be true?

A Danny did at most 130 sit-ups on Sunday.

B Danny did 25 sit-ups on Saturday.

C Danny did fewer than 190 sit-ups on Sunday.

D Danny did at least 31 sit-ups on Saturday.

5 Which of the following correctly describes the expression $\frac{1}{3}x + 14$?

F one-third more than fourteen times a number

G a third of fourteen times a number

H fourteen more than one-third times a number

J fourteen less than three of a number

6 Jim made *c* cell phone calls last week. This week, he made eleven more than six times the number of calls he made last week. Which expression represents the total number of calls he made last week and this week?

A $7c + 11$

B $c(6c + 11)$

C $11c + 7$

D $5c + 11$

7 Yesterday, Kelly spent *d* dollars on food for her restaurant. Today, she spent $(8d - 2)$ dollars on food. How much did she have to pay altogether in terms of *d*?

F She paid a total of two dollars less than sixteen times the amount of money she paid yesterday.

G She paid a total of two dollars more than nine times the amount of money she paid yesterday.

H She paid a total of three dollars more than nine times the amount of money she paid yesterday.

J She paid a total of two dollars less than nine times the amount of money she paid yesterday.

8 *Part A*

Match each of the algebraic sentences with the correct verbal translations from below. Write the number of the verbal translation beside the algebraic sentence.

$\dfrac{(3x - 12)}{185} = 32x - 10$ _____

$3(185x - 10) = \dfrac{-12}{32x}$ _____

$12x - 10 + 3x = 185x - 32$ _____

$12 - 10x > 185(3x + 32)$ _____

I   Three times the difference of one hundred eighty-five times a number and ten is the same as negative twelve divided by the product of thirty-two and the original number.

II   Twelve less ten times a number is more than one hundred eighty-five times the sum of the original number tripled and thirty-two.

III   Twelve times a number less ten plus three times the original number equals the difference of one hundred eighty-five times the original number and thirty-two.

IV   Three times a number is decreased by twelve, and the difference is divided by one hundred eighty-five. This quotient is equal to the difference of thirty-two times the original number and ten.

Part B

The sum of a number and 1 is at least 3 and at most 3. Write two algebraic sentences to describe the number and find the number.

_____

_____

LESSON 5.3 Interpreting Graphs

New York Performance Indicators

8.A.3 Describe a situation involving relationships that matches a given graph

8.PS.6 Represent problem situations verbally, numerically, algebraically, and graphically

8.CM.11 Draw conclusions about mathematical ideas through decoding, comprehension, and interpretation of mathematical visuals, symbols, and technical writing

VOCABULARY

A **coordinate plane** is formed when two number lines called axes intersect at right angles at their zero points.

On a graph, a **variable** represents a quantity or amount that changes. One variable is plotted on the horizontal axis, and the second variable is plotted on the vertical axis.

REVIEW

Understanding Graphs

On a coordinate plane, a graph shows a relationship between two variables.

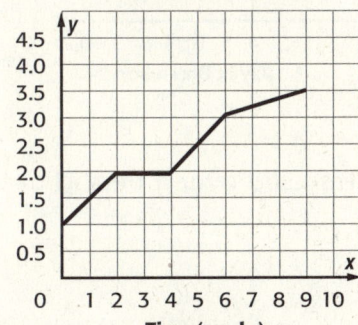

Growth of a Plant

Time (weeks)

The graph shows how one variable, height, changes as the other variable, time, increases.

What You Should Know

• The title of the graph tells you what the graph is about.

• One set of numbers (or variable) is plotted along the horizontal or *x*-axis.

• The second set of numbers (or variable) is plotted along the vertical or *y*-axis.

• Each point on a graph can be identified as an ordered pair of numbers, (x, y).

Shape of a Graph

The shape of a graph shows how two variables or quantities are related. You can identify whether a quantity increases, decreases, or stays the same as the other variable increases.

Rochester Daily High Temperatures

EXAMPLE 1

The graph to the right shows the daily high temperatures in Rochester in the first week of August. Describe how the temperature changes in this week.

You can solve this problem by studying the shape of the graph.

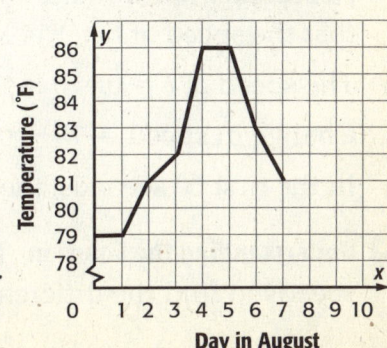

Day in August

When the temperature increases from one day to the next day, the graph line slopes up.

↑ The daily high temperature increases from day 1 to day 4.

When the temperature does not change from day to day, the graph line is flat or horizontal.

→ The daily high temperature doesn't change from day 4 to day 5.

When the temperature decreases from day to day, the graph line slopes down.

↓ The daily high temperature decreases from day 5 to day 7.

▶ **Understanding the Solution** This graph only plots points for the highest temperature on each day. The points between day 1 and day 2 on the *x*-axis don't represent the time between day 1 and day 2.

TRY IT!

The graph to the right shows the stock price of a pharmaceutical company in the first week of December. What is the greatest change in the price of the stock between any two consecutive days?

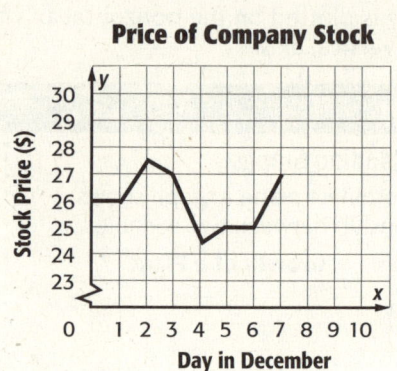

Price of Company Stock

The speed of a moving object or the distance it has traveled over a period of time can be shown graphically.

EXAMPLE 2

The graph to the right shows the speed of a car in the first 5 minutes of its travel time. How much did the speed increase in the first 60 seconds?

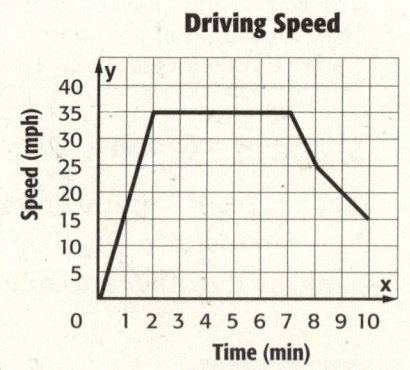

Driving Speed

You can solve this problem by finding the speed at $t = 0$ and $t = 60$ seconds.

The time in the graph is shown in minutes.

60 seconds = 1 minute, and the point (1, 15) shows that the speed at $t = 1$ is 15 miles per hour or 15 mph.

The speed at $t = 0$ is 0 mph.

Change in speed = $15 - 0 = 15$

In the first 60 seconds, the speed increased by 15 mph.

▶ **Understanding the Solution** To find the change in speed, you must subtract the two speeds to find the difference.

TRY IT!

The graph to the right shows distance covered by a runner during a 10-km race. How much distance did the runner cover between hour 2 and hour 4?

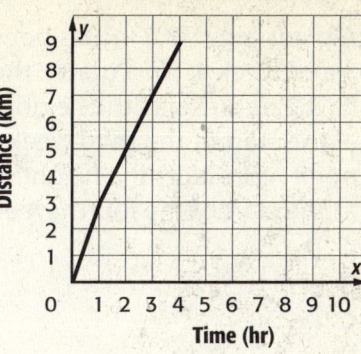

Exercises

SHORT RESPONSE

1 Billy plays marbles against his friends every day for ten days. The graph below shows how many marbles he wins or loses on each day that he plays. If Billy starts out owning 45 marbles, how many marbles does he own after the ten days are over?

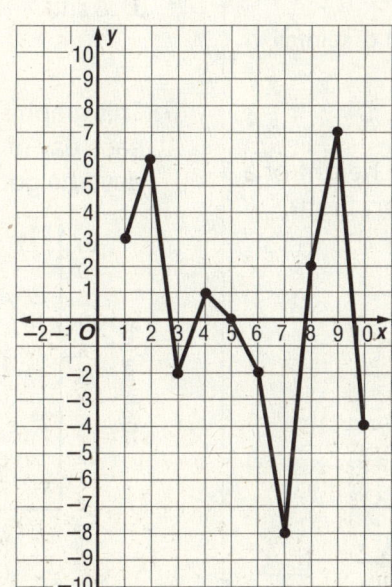

x = Day
y = Marbles Won or Lost

Show your work.

Answer _____ marbles

2 A plant grows 3.5 inches between week 3 and week 4. On a graph, the point (3, 4.2) represents the height of the plant at the start of the third week. Which point represents the height of the plant at the start of the fourth week?

A (4, 0.7)

B (4, 7.7)

C (7.7, 4)

D (3, 7.7)

3 A line graph shows the monthly rainfall in Syracuse. If the line is flat between April and May, what must be true about the rainfall in those months?

F There was no rainfall in either month.

G There was more rainfall in April than in May.

H There was more rainfall in May than in April.

J There was the same amount of rainfall in both April and May.

4 The graph below shows the height of a hot-air balloon during a short flight.

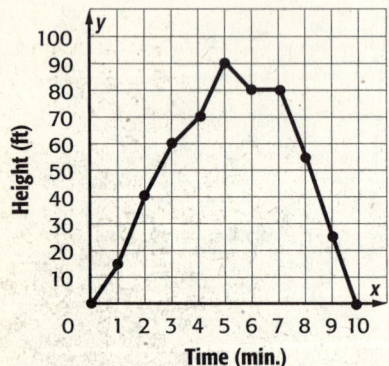

How much altitude did the balloon gain between the third and fifth minute of the flight?

A 10 feet

B 20 feet

C 30 feet

D 40 feet

5 Annie made the graph below to keep track of her height as she grows older.

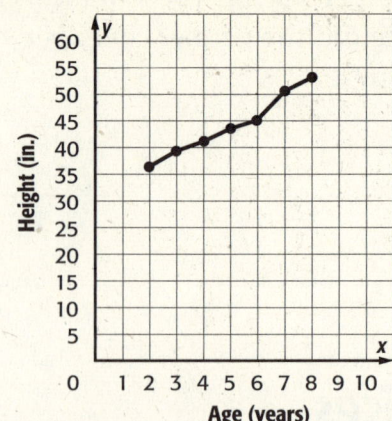

Between which consecutive years did Annie grow the most?

F 2 and 3

G 3 and 4

H 5 and 6

J 6 and 7

6 This graph shows how far Marco hiked on each day of a camping trip. On one of the days, he got a blister on his foot.

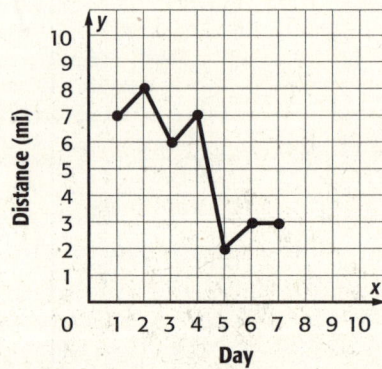

Which day is most likely the day he got the blister?

A Day 1

B Day 3

C Day 5

D Day 7

7 Patrick weighs his cat every day for a week. Every day, his cat weighs the same amount. What would the graph of the cat's weight look like?

F a horizontal line

G a vertical line

H a line sloping up

J a line sloping down

8 Christina and Maria raced against each other in the 100-meter dash. The two graphs below show how many meters each sprinter traveled in the first 10 seconds of the race.

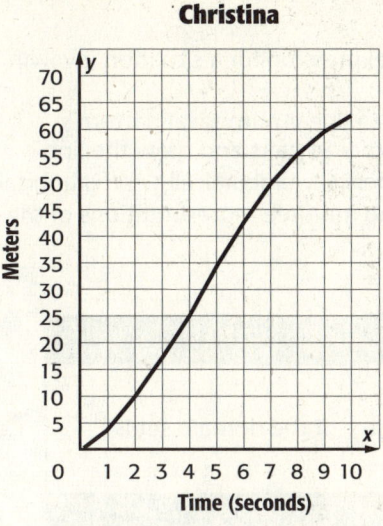

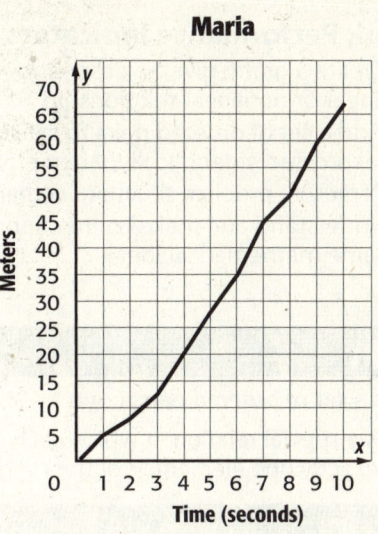

Part A

Which sprinter got off to a better start than the other? Use the graphs to find the total distance each sprinter had covered by the fifth second of the race.

Christina's distance = _____

Maria's distance = _____

Which sprinter got a better start? _____

Part B

By the tenth second of the race, which sprinter was in the lead?

Christina's distance = _____   **Maria's distance** = _____

Answer _____

Part C

Which sprinter do you think won the race? Explain your answer.

LESSON 5.4 Drawing Linear and Nonlinear Graphs

New York Performance Indicators

8.A.4 Create a graph given a description or an expression for a situation involving a linear or nonlinear relationship

8.A.16 Find a set of ordered pairs to satisfy a given linear numerical pattern (expressed algebraically); then plot the ordered pairs and draw the line

8.PS.6 Represent problem situations verbally, numerically, algebraically, and graphically

8.CN.1 Understand and make connections among multiple representations of the same mathematical idea

VOCABULARY

A **relation** is a set of ordered pairs (x, y).

A **function** is a special relation in which each element, x, of the domain set is paired with exactly one element, y, of the range set.

REVIEW

Understanding Graphs

You can describe a relation by listing its x- and y-values in a table. Its ordered pairs can also be written as a set, as shown below.

{(1, 2), (2, 3), (3, 0), (4, 5)}

You can draw a graph of a relation by plotting all the ordered pairs.

What You Should Know

If the graph of a relation is a straight line, then the relationship between x and y is linear.

If the graph of a relation is not a straight line, then the relationship between x and y is nonlinear.

Graphing Linear Relationships

The graph of a linear equation is a straight line.

x	y = 2x + 1
0	1
1	3
2	5
3	7

In a linear relationship, as the x-value changes by 1 unit, the y-value changes by a constant amount.

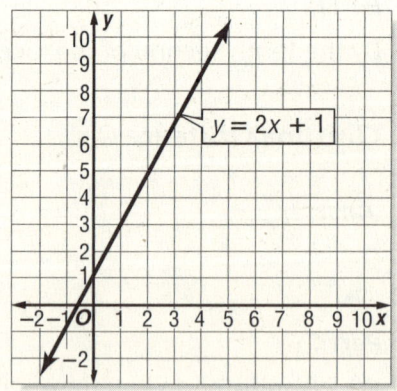

You can draw the graph of a linear equation by plotting any two points on its graph and joining them with a straight line.

When an object travels at a steady rate (constant speed), the distance-time graph is linear, or a straight line.

EXAMPLE 1

Judy is traveling from Albany to New York. After getting on the highway, she travels at a constant speed of 60 miles per hour for 3 hours. She stops for an hour to get lunch at a diner. After lunch, she drives for 2 more hours at 30 miles per hour on a country road. Draw a graph to show the total distance Judy drives.

You can solve this problem by making a table and drawing a graph

Use $d = v \times t$ to find the distance Judy travels in each hour.

Time (hours)	Distance (miles)
1	60
2	120
3	180
4	180
5	210
6	240

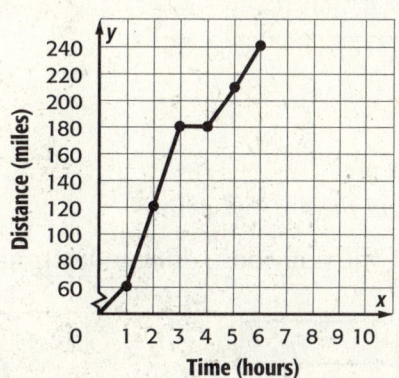

Notice that there is no change in distance between the third and the fourth hour. This is shown by the flat (horizontal) line on the distance-time graph.

In the first three hours, Judy is traveling at the speed of 60 miles per hour. This is shown by the steepest line on the graph. In the last two hours, the speed is 30 miles per hour.

▶ **Understanding the Solution** The graph is made up of three different line segments.

TRY IT!

Nadeem is training for a bicycle race. On Monday, he bicycled for 30 minutes at 30 miles per hour. He rested for 10 minutes, and then bicycled for 20 minutes at 36 miles per hour. Complete the table below and draw the distance-time graph.

Time (min.)	Distance (miles)
10	
20	
30	
40	
50	
60	

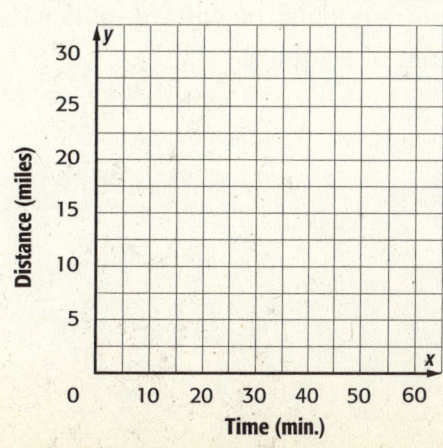

Graphing Nonlinear Relationships

The two graphs shown below represent nonlinear relationships. As the *x*-values change by 1 unit, the *y*-values do not change by a constant amount.

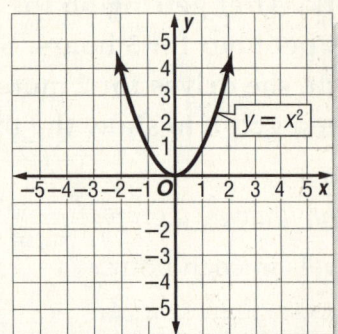
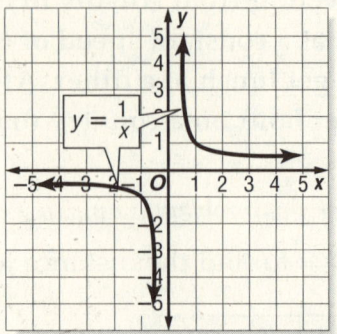

EXAMPLE 2

Graph $y = -x^2 + 2$.

Use the equation to find the ordered pairs.

To find the graph of the equation, choose both negative and positive values for *x*.

x	$y = -x^2 + 2$	(*x*, *y*)
−2	−2	(−2, −2)
−1	1	(−1, 1)
0	2	(0, 2)
1	1	(1, 1)
2	−2	(2, −2)

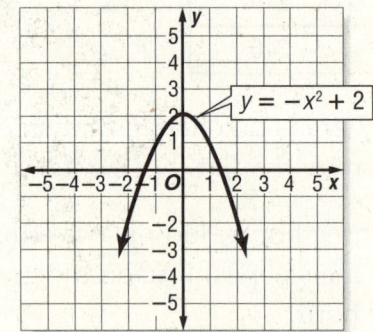

Plot the ordered pairs and connect them.

Understanding the Solution Since the graph is not a straight line, the equation $y = -x^2 + 2$ describes a nonlinear relationship.

TRY IT!

The area of a square equals the square of its sides, $A = s^2$. Draw the graph that represents the area of a square.

Exercises

SHORT RESPONSE

1 The perimeter P of a square is related to the side s of a square by $P = 4s$. Complete the table below and draw its graph.

s (cm)	P (cm)
0.5	
1	
1.5	
2	

MULTIPLE CHOICE

2 Which of the graphs below represents a linear relationship?

A
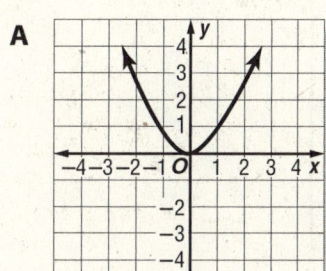

B
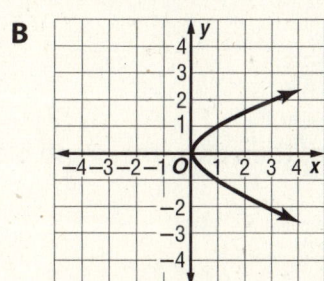

C
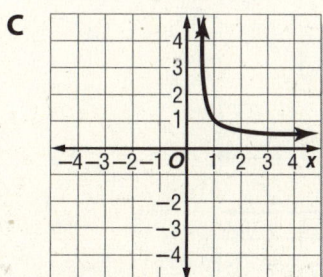

D
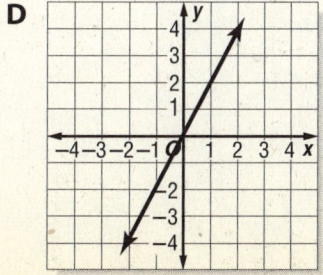

3 Which of the following relationships is nonlinear?

F $y = -x$
G $y = 2x + 1$
H $xy = 1$
J $y = -4$

4 In the equation $y = 2x - 1$, what is the change in y-values as x-values change by 1 unit?

A -1
B 0
C 1
D 2

5 Mark began graphing a linear relationship by plotting the points (1, 1) and (2, 2). What point will he plot when $x = 4$?

F (4, 3)
G (4, 0)
H (4, 4)
J (4, −4)

8 Consider the equation $y = \frac{1}{2}x^2 - 9$.

Part A

Fill in the following chart with the correct values of y.

x	$y = \frac{1}{2}x^2 - 9$
−6	
−5	
−4	
−3	
−2	
−1	
0	
1	
2	
3	
4	
5	
6	

Part B

Plot the points that you found above on the coordinate plane. Then draw the graph of the equation.

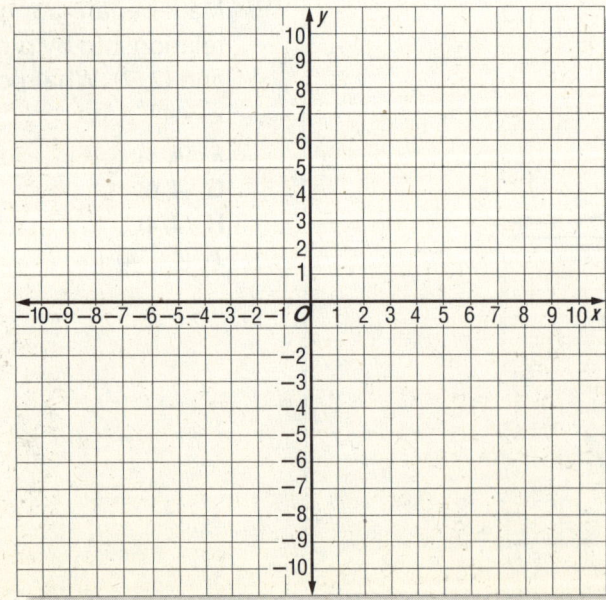

Problem-Solving Strategy: Applying Mathematics

 New York Performance Indicators

8.CN.4 Model situations mathematically, using representations to draw conclusions and formulate new situations

You can use algebraic methods to solve many different types of problems. For example, you can measure temperature using both the Fahrenheit and Celsius scales. On the Fahrenheit scale, water freezes at 32° F and boils at 212° F. On the Celsius scale, water freezes at 0° C and boils at 100° C. How can you use this information to convert between the two scales?

Example: Find a formula to convert from degrees Celsius to degrees Fahrenheit.

SOLUTION

What do you know?

$0° \text{ C} \longrightarrow 32° \text{ F}$

$100° \text{ C} \longrightarrow 212° \text{ F}$

What do you need to find?

A formula to convert between the two scales

Find a relationship.

You know that a change of 100° C corresponds to a change of 180° F (212 − 32). So a 1° change on the Celsius scale is equivalent to a change of $\frac{180}{100} = \frac{9}{5}$ degrees Fahrenheit.

Suppose the temperature is 20° C. To find the temperature in degrees Fahrenheit, multiply 20 by the factor $\frac{9}{5}$.

Since 0° C is equivalent to 32° F, you also have to add 32 to finish the conversion.

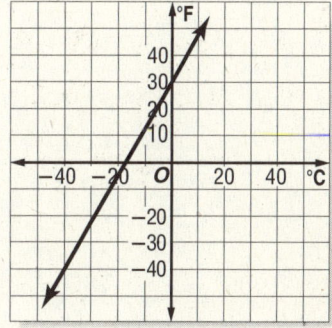

$$F = \frac{9}{5}(20) + 32 = 68$$

So the formula to convert °F to °C is

$$F = \frac{9}{5}C + 32$$

The straight-line graph shows that °F and °C have a linear relationship.

▶ **Understanding the Solution** The factor $\frac{9}{5}$ is close to 2. To find an approximate temperature in degrees Fahrenheit, you can multiply the Celsius temperature by 2 and add 32.

SHORT RESPONSE

1 Use the formula $F = \frac{9}{5}C + 32$ to find a formula for converting °F to °C.

Show your work.

Answer ____________________

2 On the lines below, use words to describe how to approximate 100°F in °C. Then do the conversion and write the approximate number of degrees Celsius.

____

____

Show your work.

Answer: 100 °F is approximately equal to ____________________.

LESSON 5.6 Using Models to Understand Polynomial Operations

New York Performance Indicators

8.A.5 Use physical models to perform operations with polynomials

8.CN.4 Model situations mathematically, using representations to draw conclusions and formulate new situations

8.R.1 Use physical objects, drawings, charts, tables, graphs, symbols, equations, or objects created using technology as representations

---

REVIEW

Understanding Polynomial Models

You can model polynomials using different tiles to represent each term.

$x^2 - 2x + 8$

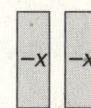 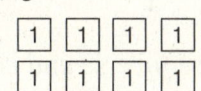

$3x - 5$

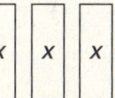 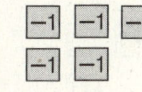

Positive tiles are white.
Negative tiles are shaded

Applying Polynomial Models

With like terms, a positive tile and a negative tile create a zero pair with a value of 0.

$x^2 + (-x^2) = 0$

$3x + (-3x) = 0$

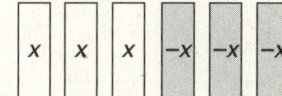

Modeling Addition and Subtraction of Polynomials

You can use models to add and subtract polynomials.

EXAMPLE 1

Add $(3x^2 + 2x + 1)$ and $(-x^2 + x - 4)$.

You can solve this problem by combining the models for the polynomials.

$3x^2 + 2x + 1$: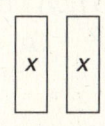

$-x^2 + x - 4$: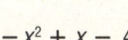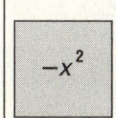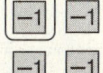

$3x^2 + (-x^2)$ $+ 2x + x$ $+ 1 + (-4)$

To combine like terms, circle any zero pairs and count the tiles left to find sums. x^2 and $-x^2$ form a zero pair. There are two x^2 tiles left, so $3x^2 + (-x^2) = 2x^2$.

There are no zero pairs in the x terms, so $2x + 1x = 3x$.

1 and -1 form a zero pair. This leaves three -1 tiles, or $1 + (-4) = -3$.

Now write a new polynomial combining the sums of the like terms.

$$2x^2 + 3x + (-3) \text{ or } 2x^2 + 3x - 3$$

▶ **Understanding the Solution** Your answer should show all of the tiles left in your model after removing the zero sums of positive and negative tiles.

TRY IT!

Use a model to add $(x^2 + 4x - 7)$ and $(-2x^2 - 3x - 4)$.

EXAMPLE 2

Subtract $(2x^2 + 2x - 3) - (-x^2 + 3x + 2)$.

You can subtract the polynomials by using the additive inverse and combining models.

In order to subtract, you need to add the additive inverse, or opposite, of the second polynomial. Rewrite the problem as $(2x^2 + 2x - 3) + (x^2 - 3x - 2)$.

Combine like terms. Remove all zero pairs to find sums. Then count the tiles.

$$2x^2 + x^2 = 3x^2$$
$$2x + (-3x) = -x$$
$$-3 + (-2) = -5$$

The sum is $3x^2 + (-x) + (-5)$ or $3x^2 - x - 5$.

▶ **Understanding the Solution** When you subtract polynomials, you are still combining like terms. Subtracting $(-x^2 + 3x + 2)$ is the same as adding $x^2 - 3x - 2$.

TRY IT!

Use a model to subtract $(x^2 + 2x + 1)$ from $(4x^2 + 3x + 2)$.

Modeling Multiplication of Polynomials

To find the area of a rectangle, you multiply the length by the width. You can model the area to see how multiplication of a polynomial and monomial works.

EXAMPLE 3

What is the area of a rectangular swimming pool that measures $3x$ feet on one side and $(x + 3)$ feet on the other side?

To find out what tiles are needed to fill a rectangle with sides of $x + 3$ and $3x$, first lay out the tiles for the horizontal side and the vertical side.

To fill the space between a horizontal x-tile and a vertical x-tile, you need an x^2-tile.

$$3x(x) = 3x^2$$

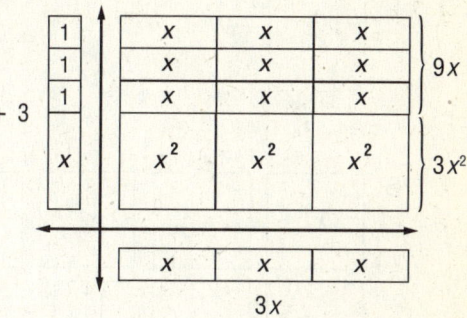

To fill the space between a horizontal x-tile and a vertical 1-tile, you need an x-tile.

$$3x(3) = 9x$$

So $3x(x + 3) = (3x^2 + 9x)$ square feet.

▶ **Understanding the Solution** When the length of a rectangle is represented by a polynomial of two terms (a binomial) and the width is represented by a monomial, the area of the rectangle represents the product of the binomial and monomial.

TRY IT!

What is the area of a rectangular piece of fabric that measures $(x + 3)$ feet long and $(x + 2)$ feet wide?

Exercises

SHORT RESPONSE

1 Find the difference of the following polynomials.

$$(6x^2 + 6x + 2) - (5x^2 - 7x + 4)$$

Show your work.

Answer _____

2 Which model shows the product of $2x$ and $3x + 1$?

A

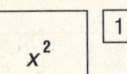

B

C

D

 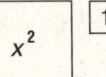

3 Which solution is correct?

$$(5a^2 - 2a + 5) - (a^2 + 2a - 5)$$

F $-4a + 6a^2 + 5$

G $4a^2 - 4a$

H $10 + 4a^2$

J $4a^2 - 4a + 10$

4 What expression represents the area of the picture frame shown below?

A $x^2 + 2x$

B $x^3 + 4x + 4$

C $x^4 + 5x$

D $2x^2 + 4x$

5 Which model does not show a zero sum?

F

G

H
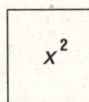

J

6 Use models to find the perimeter of each figure. Draw your models in the space provided.

Part A   *Show your work.*

$s = x + 1$

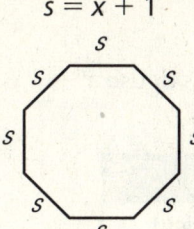

Perimeter = _____

Part B

$s = x + 2$ and $s_1 = x - 1$

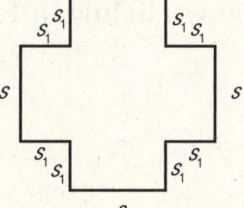

Perimeter = _____

LESSON 5.7 Adding and Subtracting Polynomials

New York Performance Indicators

8.A.7 Add and subtract polynomials (integer coefficients)

8.CN.3 Connect and apply a variety of strategies to solve problems

8.PS.8 Understand how to break a complex problem into simpler parts or use a similar problem type to solve a problem

VOCABULARY

A **numerical coefficient** is a number that is multiplied by a variable. The coefficient is usually written in front of the variables. In the monomial $3xy^2$, 3 is the coefficient.

REVIEW

Understanding Addition and Subtraction of Polynomials

You can add or subtract polynomials without using models by combining like terms. The sum or difference is the expression written in simplest form.

$$(2x + 3) + (8x - 9) = 10x - 6$$

What You Should Know

To subtract a polynomial, you can add the additive inverse or opposite of each term. Adding the opposite of a number is the same as subtracting the number.

$$12x - (-5x + 3) = 12x + (5x - 3)$$
$$= 17x - 3$$

Adding Polynomials

Using the commutative and distributive properties can often help you add polynomials.

EXAMPLE 1

The measurements of a rectangle to be fenced are shown below in feet. How many feet of fencing will need to be used?

$$8c - 3$$

$$c + 1$$

You can solve this problem by finding the perimeter of the rectangle.

Add the lengths of all four sides of the rectangle.

$$(8c - 3) + (c + 1) + (8c - 3) + (c + 1)$$

Group like terms using parentheses.

$$(8c + c + 8c + c) + (-3 + 1 - 3 + 1)$$

Use the distributive property to add the coefficients.

$$c (8 + 1 + 8 + 1) + (-3 + 1 - 3 + 1)$$

Simplify the expression and write the answer in terms of units.

$$(18c - 4) \text{ feet}$$

▶ **Understanding the Solution** After combining like terms, the sum has fewer terms than the original expression.

TRY IT!

A triangle has side lengths of $2a$, $a + 5$, and $2a + 4$. What is the perimeter of the triangle?

EXAMPLE 2

Simplify $(3y^2 + y - 10) + (9y^2 + 2y - 2)$.

You can solve this problem by adding horizontally or vertically.

Group like terms. $(3y^2 + 9y^2) + (y + 2y) + (-10 - 2)$

Use the distributive property to add the coefficients.

$$y^2 (3 + 9) + y(1 + 2) + (-10 - 2)$$

Simplify. $12y^2 + 3y - 12$

You can also add vertically. Line up the like terms and add coefficients and constants.

$$
\begin{array}{r}
3y^2 + y - 10 \\
+\ 9y^2 + 2y - 2 \\
\hline
12y^2 + 3y - 12
\end{array}
$$

▶ **Understanding the Solution** Both methods for adding polynomials give you the same sum.

TRY IT!

Simplify $(12a^2 - a - 7) + (6a^2 + 2a - 20)$.

Subtracting Polynomials

EXAMPLE 3

Find the difference: $(19y^2 + 16y - 5) - (8y^2 + 4y - 8)$.

You can subtract horizontally or vertically.

To subtract, add the opposite of each term in the second polynomial.

$$(19y^2 + 16y - 5) + (-8y^2 - 4y + 8)$$

Group like terms. $(19y^2 - 8y^2) + (16y - 4y) + (-5 + 8)$

Use the distributive property to subtract the coefficients.

$$y^2(19 - 8) + y(16 - 4) + (-5 + 8)$$

Simplify. $11y^2 + 12y + 3$

You can also subtract vertically using the additive inverse. Line up the like terms and then combine coefficients and constants.

$$19y^2 + 16y - 5$$
$$\underline{+ (-8y^2 - 4y + 8)}$$
$$11y^2 + 12y + 3$$

▶ **Understanding the Solution** Is your answer correct? Check that each term in the second trinomial has been written as its opposite. The trinomial $-(8y^2 + 4y - 8)$ should be written as $+(-8y^2 - 4y + 8)$ before grouping like terms.

TRY IT!

Simplify $(29b^2 - 16b - 12) - (20b^2 + 4b + 3)$.

Exercises

SHORT RESPONSE

1 Simplify $(25y^2 + 5y + 4) - (-9y^2 - 12y - 7)$. Show your work in the space below and explain each step.

Show your work.

Answer _____

2 Simplify $(9y^2 - 2) + (y^2 - 4)$.

A $9y^2 + 6$

B $10y^2 - 6$

C $15y^2 + 6$

D $16y^2 - 6$

3 What is the perimeter of the carpet below? Dimensions are in feet.

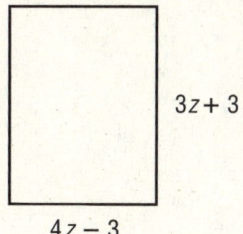

$3z + 3$

$4z - 3$

F $7z$ feet

G $10z - 6$ feet

H $13z - 12$ feet

J $14z$ feet

4 Simplify $(8x^2 + 2x + 10) + (4x^2 - 7)$.

A $12x^2 + 5$

B $12x^2 + 2x - 3$

C $12x^2 + 2x + 3$

D $16x^2 + 5x - 17$

5 What must you add to $6s^2 + 3s$ to get $5s^2 + 2s + 16$?

F $-s^2 - s + 16$

G $s^2 - s + 16$

H $s^2 + 2s - 16$

J $11s^2 + 5s + 16$

6 Simplify.

$(3m^2 + 4m + 17) - (9m^2 + 2m - 3)$

A $6m^2 + 2m + 14$

B $6m^2 + 2m - 20$

C $-3m^2 + 2m + 14$

D $-6m^2 + 2m + 20$

7 What is the sum of the missing side lengths if the perimeter of the figure below is $26a - 8$? Dimensions are in inches.

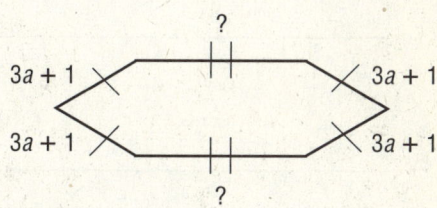

$3a + 1$ $3a + 1$

$3a + 1$ $3a + 1$

F $(26a - 4)$ inches

G $(26a - 8)$ inches

H $(14a - 12)$ inches

J $14a$ inches

EXTENDED RESPONSE

8 Simplify each of the following polynomials in two steps:

a. Group like terms using parentheses.

b. Combine like terms to simplify.

$$6abc^2 + 8a^3b - 23abc + 16a^3b - 4abc + 50abc^2 + 21abc$$

a. ______________________________

____

b. ______________________________

$$-12r^5s^2t + 8r^2st^4 + 77rs^3t^5 - 5r^2st^4 - 32rs^3t^5 + r^5s^2t + 4rs^3t^5 - 82rst$$

a. ______________________________

____

b. ______________________________

On the lines below, explain the purpose of grouping like terms before simplifying.

____

____

____

LESSON 5.8 Multiplying and Dividing Monomials and Binomials

New York Performance Indicators

8.A.6   Multiply and divide monomials

8.A.8   Multiply a binomial by a monomial or a binomial (integer coefficients)

8.PS.8 Understand how to break a complex problem into simpler parts or use a similar problem type to solve a problem

8.CM.3 Organize and accurately label work

VOCABULARY

A **binomial** is a polynomial with two terms.

A **trinomial** is a polynomial with three terms.

In the **standard form** for polynomials, terms are arranged so the powers of one variable are in descending order. For example, the trinomial $a^2 - 6a + 8$ is in standard form.

The term **FOIL** describes a method for multiplying two binomials. Using FOIL, the **F**irst terms in each binomial are multiplied, then the **O**uter, **I**nner, and **L**ast terms.

REVIEW

Understanding Multiplication and Division of Monomials

When adding or subtracting monomials with variables, the exponents stay the same.

$3x^2 + 9x^2 = 12x^2$

When multiplying or dividing monomials with variables, you operate on the exponents.

$(6x^2)(2x^3) = (6 \cdot x \cdot x)(2 \cdot x \cdot x \cdot x)$
$= (6 \cdot 2)(x \cdot x \cdot x \cdot x \cdot x)$
$= 12x^5$

$\dfrac{12x^5}{2x^3} = \dfrac{12 \cdot x \cdot x \cdot \cancel{x} \cdot \cancel{x} \cdot \cancel{x}}{2 \cdot \cancel{x} \cdot \cancel{x} \cdot \cancel{x}} = \dfrac{12 \cdot x^2}{2} = 6x^2$

What You Should Know

The laws of exponents apply when multiplying and dividing monomials.

Add exponents when multiplying.
$$(3a^2)(10a^4) = 30a^6$$

Subtract exponents when dividing.
$$\frac{28b^{10}}{4b^2} = 7b^8, b \neq 0$$

Notice that the powers being multiplied or divided have the same base.

Multiplying and Dividing Monomials

You can multiply and divide monomials with variables using the laws of exponents.

EXAMPLE 1

Simplify $(3x^2)(-6x^3)$.

You can simplify this expression by multiplying the coefficients and adding the exponents.

Use the commutative property to rearrange the factors.

$$(3x^2)(-6x^3) = (3)(-6)(x^2)(x^3)$$

Multiply the coefficients. $\quad = (-18)(x^2)(x^3)$

Add the exponents. $\quad = -18x^5$

▶ **Understanding the Solution** Is your answer correct? Check your solution by dividing $-18x^5$ by $3x^2$. Divide the coefficients and subtract the exponents to get $-6x^3$.

TRY IT!

Simplify $(3a^3)(4a^4)$.

EXAMPLE 2

Divide $\dfrac{26y^8}{13y^3}$.

You can solve this problem by dividing the coefficients and subtracting the exponents.

$$\frac{26y^8}{13y^3} = 2y^{8-3}$$

Divide the coefficients and subtract the exponents.

$$= 2y^5$$

▶ **Understanding the Solution** Is your answer correct? Check your solution by multiplying $2y^5$ by $13y^3$. You should get $26y^8$.

TRY IT!

Simplify $\dfrac{27z^{11}}{3z^2}$.

Multiplying Monomials and Binomials

To multiply a binomial by a monomial, multiply each term of the binomial by the monomial using the distributive property.

EXAMPLE 3

Simplify $(9x^2)(3x^3 - 4)$. **Write in standard form.**

You can solve this problem using the distributive property.

Rewrite the problem so each term in the binomial is multiplied by the monomial.

$$(9x^2)(3x^3 - 4) = (9x^2)(3x^3) - (9x^2)(4).$$

Multiply the monomials and write the product in standard form.

$$= 27x^5 - 36x^2$$

▶ **Understanding the Solution** Check your solution by dividing $27x^5 - 36x^2$ by $9x^2$. Divide each term by $9x^2$.

Simplify $3n^7(2n^5 - 1)$.

Multiplying Binomials

To multiply a binomial by a binomial, you must multiply each term of the first binomial by each term of the second binomial. The FOIL method helps you keep track of which terms you are multiplying.

EXAMPLE 4

Simplify $(3x - 2)(5x + 2)$. Write the product in standard form.

You can solve this problem using FOIL.

Multiply the **First** terms in each binomial. $(3x)(5x) = 15x^2$

Multiply the **Outer** terms. $(3x)(2) = 6x$

Multiply the **Inner** terms. $(-2)(5x) = -10x$

Multiply the **Last** terms. $(-2)(2) = -4$

Write the product as the sum of the FOIL products and simplify.

$$(3x - 2)(5x + 2) = 15x^2 + 6x - 10x - 4$$
$$= 15x^2 - 4x - 4$$

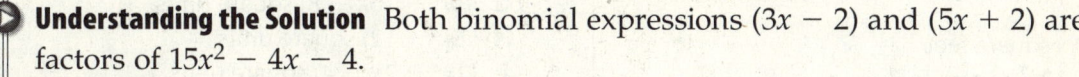 **Understanding the Solution** Both binomial expressions $(3x - 2)$ and $(5x + 2)$ are factors of $15x^2 - 4x - 4$.

TRY IT!

Simplify $(4a - 5)(2a - 5)$.

Exercises

SHORT RESPONSE

1 Simplify $(4n - 2)(5n + 6)$.

Show your work.

$(4n - 2)(5n + 6) =$ _____

2 $(3x + 1)(5x + 5)$ is equivalent to which expression?

  A $8x^2 + 5$

  B $15x^2 + 5x + 5$

  C $15x^2 + 15x + 5$

  D $15x^2 + 20x + 5$

3 The length of a rectangular poster is $(3n - 1)$ inches and the width is 5 inches. What is the area of the poster?

  F $(8n^2 - 5n)$ square inches

  G $(15n^2 - 2n)$ square inches

  H $(15n - 5)$ square inches

  J $7n$ square inches

4 If a garden plot has the dimensions shown below, what expression represents its total area?

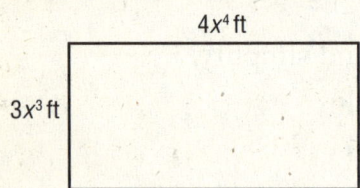

  A x square feet

  B $12x^7$ square feet

  C $7x^{12}$ square feet

  D $12x^4$ square feet

5 Which is equivalent to $(2p - 7)(3p + 8)$?

  F $6p^2 - 56$

  G $6p^2 - 21p + 15$

  H $6p^2 - 5p - 56$

  J $4p^2 + 5p - 56$

6 $2y(3y - 10)$ equals which of the following expressions?

  A $6y^2 - 10$

  B $6y^2 - 20y$

  C $-12y$

  D $-14y$

7 What is the area of the shaded portions of the figure?

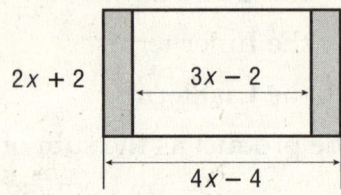

  F $(8x^2 - 8)$ square units

  G $(8x^2 - 16x - 8)$ square units

  H $(2x^2 - 12)$ square units

  J $(2x^2 - 2x - 4)$ square units

EXTENDED RESPONSE

8 Simplify both expressions below.

Show your work.

$$5xy^2z + (-10x^2z)$$

Can you use the distributive property to
simplify this expression? _____

Answer _____

$$(5xy^2z)(-10x^2z)$$

Answer _____

On the lines below, explain why finding like terms is necessary for addition and
subtraction of monomials and polynomials, but not for multiplication and division.

_____

_____

_____

_____

_____

LESSON 5.9 — Dividing Polynomials by a Monomial

New York Performance Indicators

8.A.9 Divide a polynomial by a monomial (integer coefficients)

8.PS.8 Understand how to break a complex problem into simpler parts or use a similar problem type to solve a problem

8.CM.3 Organize and accurately label work

8.CN.3 Connect and apply a variety of strategies to solve problems

---

REVIEW

Understanding Division of Polynomials

Use the distributive property and laws for exponents to divide polynomials by monomials.

Divide $\dfrac{(6a^{24} + 2a^{36})}{3a^{12}}$.

Divide each term by $3a^{12}$.

$$\frac{6a^{24}}{3a^{12}} + \frac{2a^{36}}{3a^{12}}$$

What You Should Know

To divide a monomial by a monomial, you divide the coefficients and subtract the exponents to divide variables.

Divide $\dfrac{24b^{12}}{6b^4}$.

Divide the coefficients and subtract the exponents.

$$\frac{24b^{12}}{6b^4} = 4b^{12-4} = 4b^8$$

Dividing by Multiplying by the Reciprocal

You can divide a polynomial by a monomial by multiplying by the reciprocal of the divisor.

EXAMPLE 1

Divide $20a^3 - 30a^2 - 15a$ by $5a$.

You can solve this problem by multiplying by the reciprocal of the divisor, $5a$.

Multiply by $\dfrac{1}{5a}$. $(20a^3 - 30a^2 - 15a)\dfrac{1}{5a}$

Use the distributive property. Notice that multiplying by the reciprocal is the same as dividing each term by the monomial.

$$\frac{20a^3}{5a} - \frac{30a^2}{5a} - \frac{15a}{5a}$$

Divide the coefficients and use division rules for exponents.

$$= 4a^2 - 6a^1 - 3a^0$$

Simplify and write the answer in standard form. $4a^2 - 6a - 3$

▶ **Understanding the Solution** Is your answer correct? Check your solution by multiplying $4a^2 - 6a - 3$ by $5a$. You should get $20a^3 - 30a^2 - 15a$.

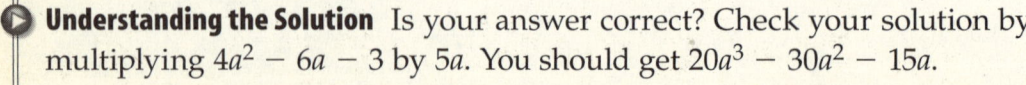

Divide $2a^3 - 12a^2 - 20a - 8$ by $2a$.

Dividing Each Term by the Divisor

Divide $12x^5 - 16x^3 - 4x - 4$ by $4x^2$.

You can solve this problem by dividing each term by $4x^2$.

Rewrite the problem as shown.

$$\frac{12x^5}{4x^2} - \frac{16x^3}{4x^2} - \frac{4x}{4x^2} - \frac{4}{4x^2}$$

Use division rules for exponents.

$$= 3x^3 - 4x^1 - \frac{1}{x} - \frac{1}{x^2}$$

Simplify.

$$= 3x^3 - 4x - \frac{1}{x} - \frac{1}{x^2}$$

▶ **Understanding the Solution** You can write the last two terms of the quotient with negative exponents: $3x^3 - 4x - x^{-1} - x^{-2}$.

Divide $15x^4 - 12x^3 - 3x - 3$ by $3x^3$.

The area of a photo frame is $48v^5 + 36v^4 + 24v^3 + 12v^2$ inches. The length is $6v$ inches. What is the width?

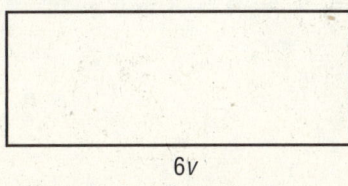

6v

You can solve the problem by dividing the area by the length.

Divide each term in the polynomial by the length, $6v$.

$$\frac{48v^5}{6v} + \frac{36v^4}{6v} + \frac{24v^3}{6v} + \frac{12v^2}{6v}$$

Use division rules for exponents. Simplify.

$$= 8v^4 + 6v^3 + 4v^2 + 2v$$

▶ **Understanding the Solution** Is your answer correct? To check your solution, multiply $8v^4 + 6v^3 + 4v^2 + 2v$ by $6v$. You should get $48v^5 + 36v^4 + 24v^3 + 12v^2$.

TRY IT!

The area of a photo frame is $96b^5 - 128b^4 + 8b^2$ inches. The width is $8b^2$ inches. What is the length?

EXAMPLE 4

Divide $(x + 2)(x - 4)$ by x^2.

You can solve this problem by writing the dividend as a polynomial.

$$(x + 2)(x - 4) = x^2 - 4x + 2x - 8$$
$$= x^2 - 2x - 8$$

Next divide each term of the polynomial by x^2.

$$\frac{(x + 2)(x - 4)}{x^2} = \frac{x^2 - 2x - 8}{x^2}$$
$$= \frac{x^2}{x^2} - \frac{2x}{x^2} - \frac{8}{x^2}$$
$$= 1 - \frac{2}{x} - \frac{8}{x^2}$$

This can also be written with negative exponents.

$$= 1 - 2x^{-1} - 8x^{-2}$$

▶ **Understanding the Solution** You can check your result by multiplying x^2 by $1 - \frac{2}{x} - \frac{8}{x^2}$. The product should equal $x^2 - 2x - 8$.

TRY IT!

Divide $(y - 3)^2$ by y.

Exercises

1 Divide $180x^{12} - 160x^8 + 40x^8 - 20x^6$ by $20x^5$. Show your work in the space below and explain each step.

Show your work.

Answer _____

2 Which expression is the quotient of $10m^3 + 6m^2 + 2m$ divided by $2m$?

A $5m^2 + 6m + 2m$

B $5m^2 + 3m + 2$

C $5m^2 + 3m + 1$

D $5m^2 + 6m$

3 The perimeter of a square is $(12s + 12)$ inches. What is the area of the square?

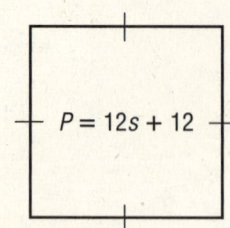

$P = 12s + 12$

F $(3s + 3)$ inches^2

G $(3s + 12)$ inches^2

H $(9s^2 + 9s + 9)$ inches^2

J $(9s^2 + 18s + 9)$ inches^2

4 Divide $9x^5 - 15x^3 - 18x^2 - 3$ by $3x^2$.

A $3x^3 - 5x - 6 - \frac{1}{x^2}$

B $3x^3 - 5x^2 - 6x - 1$

C $3x^5 - 5x^3 - 6 - \frac{1}{x^2}$

D $3x^5 - 5x^3 - 6x^2 - 1$

5 The width of a rectangular card is $5w$ inches and the area is $(10w^4 - 20w^3 - 15w)$ square inches. What is the length?

F $(5w^4 - 4w^3 - 3)$ inches

G $(2w^3 - 4w^2 - 3)$ inches

H $(2w^3 - 4w^3 - 3)$ inches

J $(2w^3 - 4w^2 + 3w)$ inches

6 $16z^8 - 20z^6 - 4z^3 - 8z^2$ divided by a monomial is $8z^5 - 10z^3 - 2 - \frac{4}{z}$. What is the monomial?

A $2z$

B $4z^2$

C z^3

D $2z^3$

7 The length and area of a rectangle are shown in the figure. What is the width?

$\ell = 5b$ in.

$A = (5b^4 - 50b)$ in^2

F $(5b^3 - 50)$ inches

G $(b^3 - 50)$ inches

H $(b^3 - 10)$ inches

J $(b - 10)$ inches

8 Divide the following polynomial by its divisor.

$$(9xy + 27x^2y^2 - 33y^2z + 51x^3z^2) \div 3xyz$$

Show your work.

Answer _____

On the lines below, explain why the answer contains some variables in the denominators.

_____

_____

_____

LESSON 5.10 — Factoring Algebraic Expressions

 New York Performance Indicators

8.A.10 Factor algebraic expressions using the GCF

8.PS.17 Evaluate the efficiency of different representations of a problem

8.CN.3 Connect and apply a variety of strategies to solve problems

8.R.4 Explain how different representations express the same relationship

VOCABULARY

The **greatest common factor**, or **GCF**, is the largest factor of a set of numbers. It is also called the **greatest common divisor**, since it is the largest number that divides into each number in a set without a remainder.

REVIEW

Understanding the GCF

You can find the GCF of 12, 18, and 42 by factoring each number completely.

$12 = 2 \times 2 \times 3$
$18 = 3 \times 2 \times 3$
$42 = 7 \times 2 \times 3$

Each set of factors has 2×3 in common, so 6 is the GCF. It is the largest number that divides into 12, 18, and 42 without remainders.

Applying the GCF to Algebraic Expressions

Find the GCF of x^4, x^3, and x.

$x^4 = x \cdot x \cdot x \cdot x$
$x^3 = x \cdot x \cdot x$
$x^7 = x \cdot x \cdot x \cdot x \cdot x \cdot x \cdot x$

Each set of factors has $x \cdot x \cdot x$ in common, so x^3 is the GCF. It is the largest expression that divides into all three monomials without a remainder.

Finding the GCF

EXAMPLE 1

Find the greatest common factor for $16x^4y^7$, $20x^8y^3$, and $8x^7y^2$.

You can solve this problem by factoring each expression completely, or using prime factorization.

$$16x^4y^7 = 2 \cdot 2 \cdot 2 \cdot 2 \cdot x \cdot x \cdot x \cdot x \cdot y \cdot y \cdot y \cdot y \cdot y \cdot y \cdot y$$
$$20x^8y^3 = 2 \cdot 2 \cdot 5 \cdot x \cdot x \cdot x \cdot x \cdot x \cdot x \cdot x \cdot x \cdot y \cdot y \cdot y$$
$$8x^7y^2 = 2 \cdot 2 \cdot 2 \cdot x \cdot x \cdot x \cdot x \cdot x \cdot x \cdot x \cdot y \cdot y$$

Now find the GCF for the coefficients 16, 20, and 8. The largest factor common to all three numbers is $2 \cdot 2 = 4$.

Find the GCF for each variable x and y separately. The largest power of x common to x^4, x^8, and x^7 is $x \cdot x \cdot x \cdot x = x^4$. The largest power of y common to y^7, y^3, and y^2 is $y \cdot y = y^2$. The GCF of the three monomial expressions is the product of 4, x^4, and y^2, or $4x^4y^2$.

▶ **Understanding the Solution**   The largest power of x common to each monomial is the smallest power of x among them, x^4. Only the smallest of the powers will divide into all three monomials. Similarly, y^2 is the smallest power of y among the three monomials.

TRY IT!

Find the GCF for $14a^8b^3$, $28a^4b^6$, and $21a^5b^5$.

Factoring Out the GCF

EXAMPLE 2

Write each of the monomials $16x^4y^7$, $20x^8y^3$, and $8x^7y^2$ as a product of the GCF $4x^4y^2$ and the remaining factors.

You can solve this problem by circling the GCF in the prime factorization of each monomial and seeing what factors are left.

In Example 1, you wrote the prime factorization of each expression and found the GCF. Now circle the GCF in each and see what the remaining factors are.

$$16x^4y^7 = \boxed{2 \cdot 2} \cdot 2 \cdot 2 \cdot \boxed{x \cdot x \cdot x \cdot x} \cdot \boxed{y \cdot y} \cdot y \cdot y \cdot y \cdot y \cdot y$$
$$20x^8y^3 = \boxed{2 \cdot 2} \cdot 5 \cdot \boxed{x \cdot x \cdot x \cdot x} \cdot x \cdot x \cdot x \cdot x \cdot \boxed{y \cdot y} \cdot y$$
$$8x^7y^2 = \boxed{2 \cdot 2} \cdot 2 \cdot \boxed{x \cdot x \cdot x \cdot x} \cdot x \cdot x \cdot x \cdot \boxed{y \cdot y}$$

Now write each expression as the product of the GCF, $4x^4y^2$, and the remaining factors.

$$16x^4y^7 = 4x^4y^2 \cdot 4y^5$$
$$20x^8y^3 = 4x^4y^2 \cdot 5x^4y$$
$$8x^7y^2 = 4x^4y^2 \cdot 2x^3$$

▶ **Understanding the Solution**   You can think of factoring as dividing by the common factor: $\dfrac{16x^4y^7}{4x^4y^2} = 4y^5$. To check your answer, multiply the GCF times the remaining factor. The product should be the original expression.

TRY IT!

Find the GCF of $14a^8b^3$, $28a^4b^6$, and $21a^5b^5$. Write each monomial as a product of the GCF and the remaining factors. Check your answer.

EXAMPLE 3

Factor the trinomial expression $20x^2y^7 + 15x^6y^9 - 25x^3y^5$.

You can solve this problem by finding the GCF for the three terms and seeing what is left from each term.

The GCF for the coefficients 20, 15, and -25 is 5. The greatest power of x that divides each term is x^2. The greatest power of y that divides each term is y^5. So the GCF is $5x^2y^5$.

Now factor out the GCF and circle what's left, or you can divide each term by the GCF.

$$\frac{20x^2y^7}{5x^2y^5} = 4y^2, \quad \frac{15x^6y^9}{5x^2y^5} = 3x^4y^4, \text{ and } \frac{-25x^3y^5}{5x^2y^5} = -5x$$

Rewrite the trinomial by writing the GCF and then writing the remaining factors of each term in parentheses. Make sure that the addition and subtraction signs in the original expression are also included in the factored expression.

$$20x^2y^7 + 15x^6y^9 - 25x^3y^5 = 5x^2y^5 (4y^2 + 3x^4y^4 - 5x)$$

▶ **Understanding the Solution**   You can check that your factoring is correct by using the distributive property and multiplying. Remember that when you multiply expressions with the same base, you multiply the coefficients and add the exponents.

TRY IT!

Factor the trinomial expression $16a^5b^3 - 24a^4b^2 - 8a^3b^7$.

Exercises

SHORT RESPONSE

1 The multiplication sentence $2 \times 6 = 12$ can be written as the division sentence $12 \div 6 = 2$. Rewrite the expression $9x^2y + 6xy^2 + 12x^3y^5$ as a multiplication sentence and as a division sentence, using the GCF of the terms.

Multiplication sentence:

_____

Division sentence:

_____

2 Find the GCF for the following terms.

$$30a^4b^7, 42a^2b^5, \text{ and } 12a^2b^8$$

A $6ab^5$

B $6a^2b^5$

C $12a^2b^5$

D $12a^4b^8$

3 Factor the binomial $10a^2b + 15$ by finding the GCF.

F $5(2a^2b + 3)$

G $5(2a^2b3)$

H $5a(2ab + 3)$

J $10a(ab + 15)$

4 Factor the expression.

$$8x^2 + 16x^4 + 32x^5$$

A $8x^2(1 + 2x^2 + 4x^3)$

B $8x^2(0 + 2x^2 + 4x^3)$

C $8x^2(2x + 4x^3 + 8x^4)$

D $4x^2(2 + 4x^2 + 8x^3)$

5 If $5x^2y^2$ is factored from the term $20x^4y^{10}$, what term remains?

F $15x^2y^8$

G $4x^2y^5$

H $4x^2y^8$

J $15x^2y^5$

6 Factor the following trinomial by finding the GCF.

$$24a - 36b + 60a^2$$

A $6a(4 - 6b + 10a)$

B $12a(4 - 3b + 5a)$

C $6(4a - 6b + 10a^2)$

D $12(2a - 3b + 5a^2)$

7 Factor the following binomial by finding the GCF.

$$24x^2y^7 + 12x^5y^8$$

F $6x^3y^6(4xy + 2x^2y^2)$

G $6x^4y^5(4y^2 + 2xy^3)$

H $12x^2y^7(2 + x^3y)$

J $24(x^4y^7 + x^5y^8)$

8 $16x^2(x + 4)$ can be written as a product of $8(x + 4)$ and what term?

A $2x^2(x + 4)$

B $16x(x + 4)$

C $128x^2(x + 4)^2$

D $2x^2$

9 Factor $3(x - 2) + 5x(x - 2)$ using the greatest common factor $(x - 2)$.

F $3x - 6 + 5x^2 - 10x$

G $-6 + 5x^2 - 7x$

H $6 + 5x$

J $(3 + 5x)(x - 2)$

10 Finding the GCF of polynomials is often used to simplify algebraic equations before solving them.

Part A

Factor out the GCF of each polynomial in the equation below. Simplify any like terms.

$$5xy^2 + 15x^2y^2 - 25xy^2 + 65x^2y^2 = 10x^2y^2 + 45xy^2$$

Show your work.

Answer _____

Part B

Factor out the GCF of each polynomial in the equation below. Simplify any like terms.

$$6a^2b^2 - 18ab^2 + 54a^2b^2 = 30ab^2 + 48a^2b^2 - 66ab^2$$

Show your work.

Answer _____

Part C

Explain how you would solve the above equations.

_____

_____

_____

LESSON 5.11 — Problem-Solving Strategy: Making Generalizations

New York Performance Indicators
8.PS.4 Observe patterns and formulate generalizations
8.RP.2 Use mathematical strategies to reach a conclusion

Observing patterns is a useful strategy for both solving problems and understanding general mathematical properties. For example, look at the pattern in the following sums of even numbers.

$2 + 4 = 6$
$4 + 8 = 12$
$6 + 10 = 16$
$14 + 12 = 26$

General Result
The sum of two even numbers is always an even number.

Problem: Find a general pattern in the sum of two odd numbers.

SOLUTION

What do you know?
Odd numbers are of the form $2k + 1$, where k is an integer.

Even numbers are multiples of 2. They are of the form $2k$, where k is an integer.

What do you need to find?
A general pattern for the sum of two odd numbers

Look for a general pattern.

$1 + 3 = 4$
$3 + 5 = 8$
$5 + 7 = 12$
$11 + 19 = 30$

The sum of two odd numbers is always an even number.

Understanding the pattern.

You can use an algebraic expression to show the general result you found.

Write expressions to represent two odd numbers. $\longrightarrow$ $2n + 1$ and $2m + 1$

Find the sum. $\longrightarrow$ $(2n + 1) + (2m + 1)$

Factor out the GCF, 2. $\longrightarrow$ $= 2n + 2m + 2$
$= 2(n + m + 1)$

▶ **Understanding the Solution** The expression $2(n + m + 1)$ is a multiple of 2, so it is always an even number. This result shows that the sum of two odd numbers is always an even number.

SHORT RESPONSE

1 Find a general pattern for the sum of an even number and an odd number. Write the pattern in words on the lines below.
Show your work.

Answer ______________________________________

____

2 An odd number can be written as $2m + 1$. An even number can be written as $2n$. Explain whether the product of an even number and an odd number is odd or even.
Show your work.

Answer ______________________________________

____

LESSON 5.12 Factoring Trinomials

New York Performance Indicators

8.A.11 Factor a trinomial in the form $ax^2 + bx + c$; $a = 1$ and c having no more than 3 sets of factors

8.PS.13 Set expectations and limits for possible solutions

8.R.7 Investigate relationships between different representations and their impact on a given problem

VOCABULARY

A **binomial** is an algebraic expression with 2 terms

A **trinomial** is an algebraic expression with 3 terms.

REVIEW

Understanding Trinomials

When you multiply two binomials, the result often simplifies to a trinomial.

Example:

$(x - 5)(x + 3) = x^2 + 3x - 5x - 15$
$(x - 5)(x + 3) = x^2 - 2x - 15$

Thus, a trinomial can often be factored into two binomial factors. The factors of $x^2 - 2x - 15$ are $(x - 5)(x + 3)$.

What You Should Know

There is a pattern to multiplying binomials that will help you to factor trinomials.

$(x - \mathbf{5})(x + \mathbf{3}) = x^2 - \mathbf{2}x - \mathbf{15}$
$-\mathbf{5} + \mathbf{3} = -\mathbf{2}$
$-\mathbf{5} \times \mathbf{3} = -\mathbf{15}$

The **sum** of the constants in the binomials gives you the middle coefficient of the trinomial, and their **product** gives you the constant in the trinomial. This pattern only works if the coefficient of the x^2-term in the trinomial is 1.

Factoring Trinomials

EXAMPLE 1

Factor the trinomial $x^2 - 3x - 18$.

You can solve this problem using the pattern for a trinomial and its binomial factors.

Start by listing pairs of factors of –18.

$$1 \text{ and } -18 \quad 2 \text{ and } -9 \quad 3 \text{ and } -6$$
$$-1 \text{ and } 18 \quad -2 \text{ and } 9 \quad -3 \text{ and } 6$$

Which of these pairs can also be added to get –3? There is only one: $3 + (-6) = -3$. Thus, the factors of $x^2 - 3x - 18$ are $(x + 3)(x - 6)$.

▶ **Understanding the Solution** You can check your result by multiplying the binomials: $(x + 3)(x - 6) = x^2 - 6x + 3x - 18 = x^2 - 3x - 18$. Notice that when the trinomial has a **negative** constant, **one** of the two constants in the binomial factors must be negative.

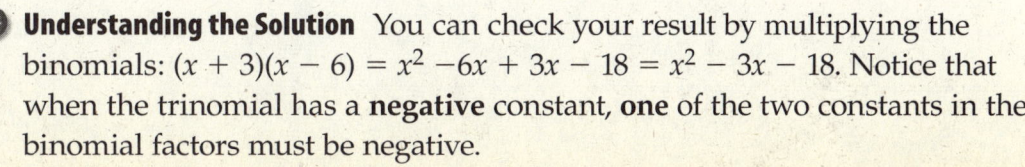

Factor the trinomial $x^2 + 2x - 8$. Check your answer by multiplying.

EXAMPLE 2

Charlie has a small square garden. One summer, he adds 4 feet to the length of the garden and 3 feet to the width, so that he can plant new flowers. Show how the area in each part of the garden can be modeled using a trinomial and its factors.

Since the size of the original square garden is unknown, you can solve this problem by calling each side of the original square garden x.

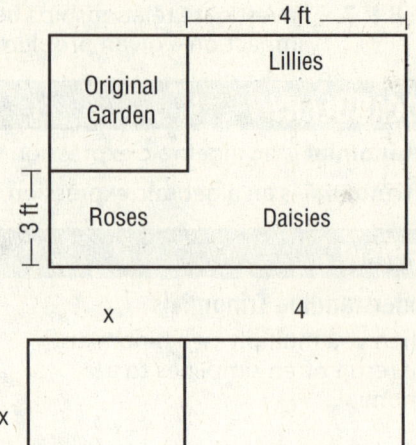

The area of the original garden is $x \cdot x = x^2$. The area for the lilies is $4 \cdot x = 4x$. The area for the roses is $3 \cdot x = 3x$, and the area for the daisies is $4 \cdot 3 = 12$.

The total area for the new garden is $x^2 + 3x + 4x + 12 = x^2 + 7x + 12$, and the sides have lengths $x + 4$ and $x + 3$.

▶ **Understanding the Solution** The area of a rectangle can be found by multiplying the length times the width. In this case, the length was $x + 4$ and the width was $x + 3$. $(x + 4)(x + 3) = x^2 + 7x + 12$. The parts of the area of the rectangle correspond to the terms in the trinomial.

TRY IT!

A patio is being enlarged to be 5 feet wider and 6 feet longer. Draw a picture to show how the area in each part of the patio can be modeled using a trinomial and its factors.

Exercises

1 Factor the trinomial $x^2 - 2x + 1$.

Show your work.

Answer _____

2 Factor $5x^4 - 15x^3 - 50x^2$ by first factoring out the greatest common factor. Then factor the remaining trinomial. Show each step.

Show your work.

Answer _____

3 Factor the following trinomial. The first factor has been given to you.

$$x^2 + 5x - 6 = (x - 1)(?)$$

A $(x - 1)(x - 6)$
B $(x - 1)(x + 6)$
C $(x - 1)(x + 5)$
D $(x - 1)(x - 5)$

4 Factor the following trinomial. The first factor has been given to you.

$$x^2 - 5x + 6 = (x - 2)(?)$$

F $(x - 2)(x - 3)$
G $(x - 2)(x + 3)$
H $(x - 1)(x + 5)$
J $(x - 1)(x - 6)$

5 Factor $x^2 + 5x + 6$.

A $(x + 1)(x + 5)$
B $(x - 2)(x + 3)$
C $(x + 2)(x + 3)$
D $(x - 2)(x - 3)$

6 Factor $x^2 - 7x + 12$.

F $(x - 2)(x - 6)$
G $(x - 4)(x - 3)$
H $(x - 12)(x + 1)$
J $(x - 3)(x + 4)$

7 Factor $x^2 + 2x + 1$.

A $(x + 1)(x + 2)$
B $(x - 2)(x + 1)$
C $(x + 1)(x + 1)$
D $(x - 1)(x - 1)$

8 Factor $x^2 - 6x + 9$.

F $(x + 3)(x + 3)$
G $(x - 2)(x + 3)$
H $(x - 3)(x - 3)$
J $(x + 1)(x - 9)$

9 An architect created the new floor plan shown below. Use factoring and the area of each new room to find out how the dimensions of the original space were changed.

Old room $A = x^2$ ft^2	New room $A = 4x$ ft^2
New room $A = 2x$ ft^2	New room $A = 8$ ft^2

A 2 feet were added to the width and 4 feet to the length.
B 1 foot was added to the width and 6 feet to the length.
C 1 foot was added to the width and 5 feet to the length.
D 2 feet were subtracted from the width and 4 feet were added to the length.

EXTENDED RESPONSE

10 Look at the following expressions.

$$5x^2y^2 - 15xy^2 - 50y^2$$
$$8a^2b^2c - 16ab^2c - 192b^2c$$

Part A

On the lines below, explain how you can factor complex polynomials completely.

___

___

___

Part B

Now factor each polynomial completely.

$5x^2y^2 - 15xy^2 - 50y^2$	$8a^2b^2c - 16ab^2c - 192b^2c$
Show your work.	***Show your work.***
Answer ___	***Answer*** ___

LESSON 5.13   Representing Numerical Information in Multiple Ways

New York Performance Indicators

8.A.15 Understand that numerical information can be represented in multiple ways: arithmetically, algebraically and graphically

8.PS.6 Represent problem situations verbally, numerically, algebraically, and graphically

8.R.7 Investigate relationships between different representations and their impact on a given problem

VOCABULARY

A **quadratic function** is of the form $y = ax^2 + bx + c$, where a is not zero. The graph of a quadratic function is a curve called a **parabola**.

REVIEW

Understanding How Objects Fall

People once thought that an object pushed forward off a high place would drop straight down. Modern physics shows that because the object continues to have forward momentum as it falls, the path it takes is a parabola.

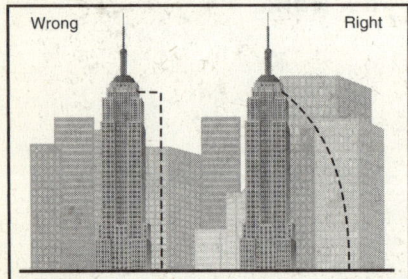

What You Should Know

You can model the height of a falling object using a quadratic function. You can find the height of the object at different times using the equation, the graph, or a table of values.

The path a falling object takes depends on factors such as the initial height of the object, the initial speed at which the object is thrown, and gravity.

Exploring the Path of a Falling Object

EXAMPLE 1

The height of a ball thrown into the air is given by the quadratic function $y = -16t^2 + 64t$, where y is the height of the ball in feet and t is time in seconds. In function notation, we write $f(t) = -16t^2 + 64t$. Find the height of the ball after 3 seconds, the time when the ball is at its highest, and how long it takes for the ball to hit the ground.

You can solve this problem by making a table of values and a graph.

Begin with inputs that make sense in the problem. Start with 0 seconds for t, then 1 second, 2 seconds, etc.

At $t = 0$, $y = -16 \times 0^2 + 64 \times 0$, or $f(0) = 0$. For an input of 0, the output is 0.

t	$y = -16t^2 + 64t$
0	0
1	48
2	64
3	48
4	0

Each row on the table corresponds to a point that can be graphed.

Graph the ordered pairs with the time or input values along the x-axis. A height of -80 does not make sense for a falling ball, so you do not have to plot the pair $(5, -80)$.

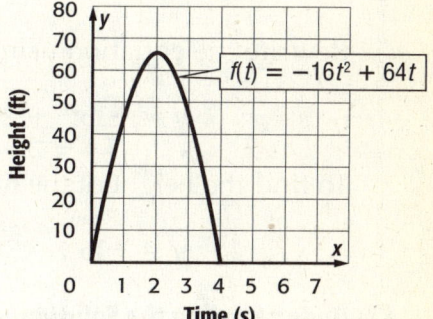

Both the graph and the table show that at 3 seconds, the ball is at a height of 48 feet, or $f(3) = 48$.

The highest point is reached after 2 seconds, a height of 64 feet.

To find how long it takes the ball to hit the ground, you can use the table to find which value for t gives a height of 0. You can also look at the graph to see where the parabola hits the x-axis. $f(4) = 0$, so the ball hits the ground after 4 seconds.

▶ **Understanding the Solution**   Many values on the table appear twice because the parabola is symmetrical. For example, the ball is 48 feet off the ground after 1 second and again after 3 seconds, on its way back down.

TRY IT!

A video game designer uses the equation $y = -2x^2 + 16x$ to model the height of a ball. Use a table to find the height of the ball at 3 seconds, the maximum height of the ball, and how long it takes the ball to hit the ground. Write your answers both in sentence form and function notation.

x	$y = -2x^2 + 16x$	x	$y = -2x^2 + 16x$
0		5	
1		6	
2		7	
3		8	
4		9	

EXAMPLE 2

A fireworks display is launched into the air from an initial height of 50 feet, at an initial vertical velocity of 200 feet per second. Write an equation for the vertical height of the ball. Use the following formula from physics: an object's height, y, is determined by the equation $y = -16t^2 +$ (initial velocity)$t +$ (initial height). Use your equation to predict the height of the fireworks after 4 seconds.

You can solve this problem by substituting the initial height and initial velocity into the equation.

Rewrite the equation using the given initial height and velocity.

$$y = -16t^2 + 200t + 50$$

To find the height of the fireworks after 4 seconds, substitute $t = 4$ in the equation.

$$y = -16 \times 4^2 + 200 \times 4 + 50 = 594 \text{ feet}$$

▶ **Understanding the Solution** An object launched with a high initial velocity will travel a large distance in a short amount of time. If the initial velocity was much slower, the object would be less far off the ground after 4 seconds.

TRY IT!

A baseball player throws a ball from an initial height of 6 feet (height of player), at an initial vertical velocity of 85 feet per second. Use $y = -16t^2 +$ (initial velocity)t + (initial height) to write an equation to determine the height of the ball. Use your equation to find the height of the ball after 3 seconds.

Exercises

SHORT RESPONSE

1 The graph shows the path of a ball. Circle the highest and lowest points on the graph. Then use the lines below to explain what these points tell you about the path of the ball.

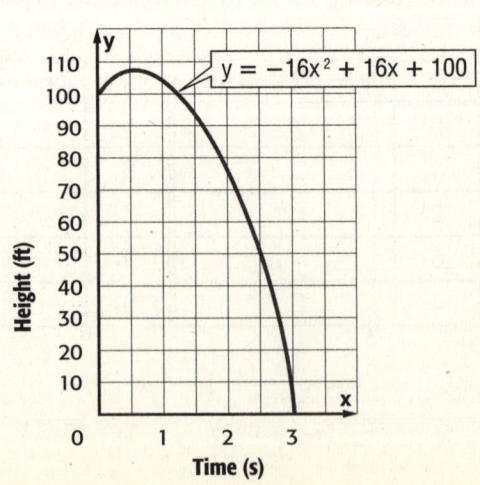

$y = -16x^2 + 16x + 100$

Height (ft)

Time (s)

2 Which of the following are quadratic functions?

1. $y = 8t^2 + 200t$
2. $f(x) = 5x^2 + 2$
3. $y = 8t + 20$

A 1 only
B 1 and 2
C 1 and 3
D 2 and 3

3 Use the table to determine the maximum height of an arrow shot in the air with an initial velocity of 128 feet per second.

t (seconds)	y (feet)
0	16
1	112
2	144
3	112
4	16

F 16
G 840
H 128
J 144

4 While mountain climbing, Cristy knocks a rock loose at a height of 1,000 feet. The rock's height after t seconds is given by $y = -16t^2 + 1,000$. What is the rock's height after 5 seconds?

A 600 ft
B 840 ft
C 1,800 ft
D 26,600 ft

Use the graph below to answer questions 5–8.

Path of a Falling Rock

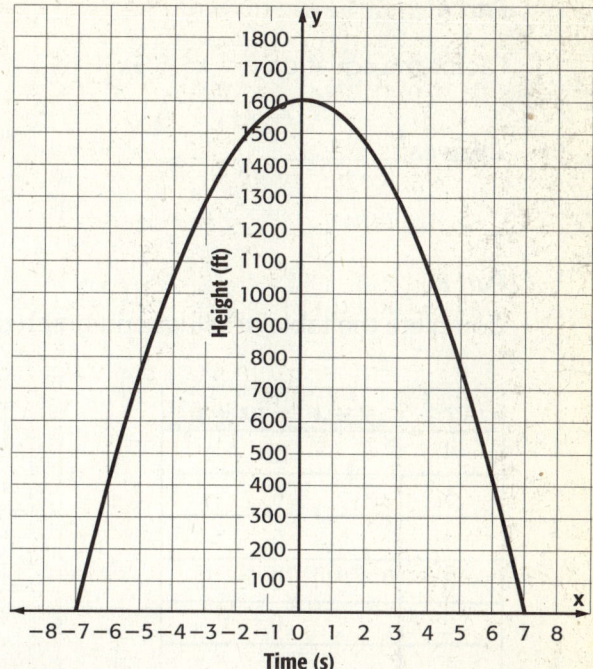

5 When does the rock hit the ground?

F $t = -7$ s
G $t = 0$ s
H $t = 5$ s
J $t = 7$ s

6 The graph shows that $f(3) = 1,300$. What does this information reveal about the path of the rock?

A At 3 seconds, the rock hits the mountain.
B At 3 seconds, the height of the rock is 1,300 feet.
C At 3 feet, the rock has traveled for 1,312 seconds down the mountain.
D The solution is 437.

7 Which statement best explains why the left side of the parabola can be ignored?

F There are two times when the rock hits the ground.
G A parabola is symmetrical.
H You cannot have negative height.
J You cannot have negative time.

8 The quadratic equation $y = x^2 + 5x + 6$ has two solutions, $x = -2$ and $x = -3$.

Part A

Factor the equation $y = x^2 + 5x + 6$.

Answer _____

Part B

Complete the table of values and graph the equation.

x	$y = x^2 + 5x + 6$
−6	12
−5	6
−4	2
−3	0
−2	0
−1	2
0	6
1	12

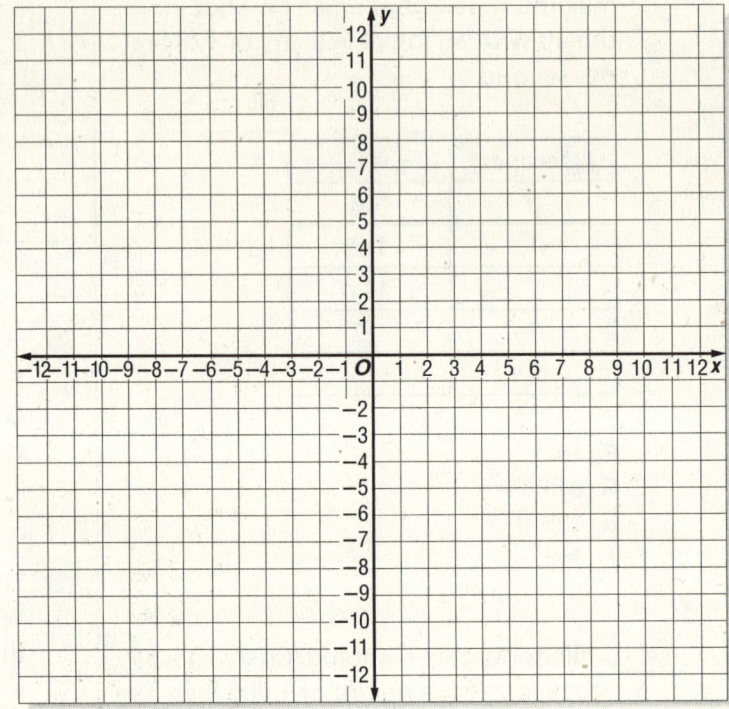

Part C

Do you see a pattern? Describe the relationship between the solutions to the equation, the factors of the equation, and the graph of the equation. Write your answer on the lines below.

MULTIPLE CHOICE

1 Chris wants to ride on the Collosal Coaster at the amusement park. Admission to the park costs $6, and each ride on the roller coaster costs another $0.50. Chris wants to spend less than $12.50 in all. Which inequality describes the number of times Chris can ride on the roller coaster?

A $6 + (0.50)x > 12.50$

B $6 + (0.50)x < 12.50$

C $6 + (0.50)x \leq 12.50$

D $6 - (0.50)x > 12.50$

2 Factor the trinomial $z^2 - 2z - 8$.

F $(z - 4)(z - 2)$

G $(z - 8)(z + 1)$

H $(z + 4)(z - 2)$

J $(z - 4)(z + 2)$

3 Which expression represents the area of the swimming pool shown below?

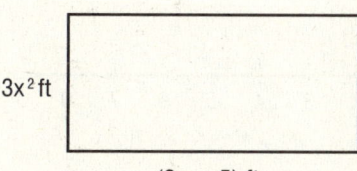

$3x^2$ ft

$(2x - 5)$ ft

A $(3y^2 + 2y - 5)$ square feet

B $(6y^2 + 4y - 10)$ square feet

C $(6y^3 - 15y^2)$ square feet

D $(12y^3 - 30y^5)$ square feet

4 Factor the binomial $12x^2y^2 + 8xy^5$ by finding the GCF.

F $4xy^2(3x + 2y^3)$

G $6xy^2(2xy + y^3)$

H $2xy^3(6x + 4y^2)$

J $4xy(3xy^2 + 2y^4)$

5 Which equation describes the following line?

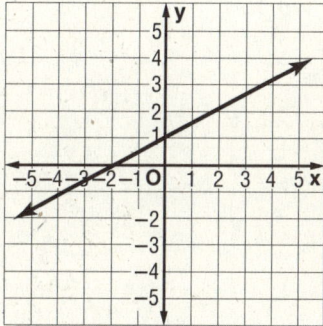

A $y = \frac{1}{2}x + 1$

B $y = x - \frac{1}{2}$

C $y = -\frac{1}{2}x + 1$

D $y = -\frac{1}{2}x - 1$

6 What is the product of $(a^2 + 2)$ and $(4a^2 - 7)$?

F $4a^4 + a - 14$

G $4a^4 + a^2 - 14$

H $4a^4 - a - 14$

J $4a^4 - a^2 - 14$

7 A line is described by the equation $y = -2x + 1$. Graph this line on the coordinate grid below. At what value of x does $y = 999$?

Show your work.

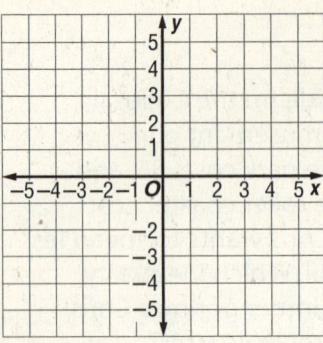

$x =$ _____

8 Find the solution to the inequality $x - 12 > -7$.

Show your work.

Answer _____

Graph your solution on the number line below.

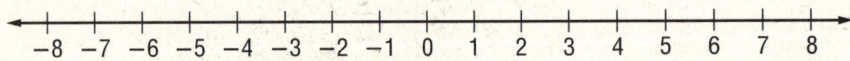

LESSON 6.1 Identifying Vertical Angles

New York Performance Indicators

8.G.1	Identify pairs of vertical angles as congruent
8.G.6	Calculate the missing angle measurements when given two intersecting lines and an angle
8.PS.16	Justify solution methods through logical argument
8.RP.2	Use mathematical strategies to reach a conclusion
8.CM.11	Draw conclusions about mathematical ideas through decoding, comprehension, and interpretation of mathematical visuals, symbols, and technical writing

VOCABULARY

Adjacent angles are two angles with the same vertex and a common side between them.

A **linear pair** of angles is formed by two adjacent angles whose noncommon sides are opposite rays.

Congruent angles have the same measure.

Intersecting lines cross each other at one and only one point.

Vertical angles are opposite angles formed by intersecting lines, and are equal in measure, or congruent.

REVIEW

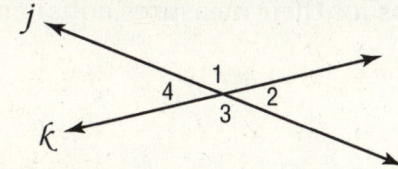

Understanding Vertical Angles

Lines j and k intersect at one point and form two pairs of vertical angles.

$\angle 1$ and $\angle 3$

$\angle 2$ and $\angle 4$

What You Should Know

The measure of angle a is written as $m\angle a$. The symbol $\cong$ means "is congruent to."

The Vertical Angles Theorem states that if two angles are vertical angles, then they are congruent.

$\angle 1 \cong \angle 3$, or $m\angle 1 = m\angle 3$

$\angle 2 \cong \angle 4$, or $m\angle 2 = m\angle 4$

The sum of the measures of a linear pair of angles is equal to 180°.

$$m\angle 1 + m\angle 2 = 180°$$
$$m\angle 2 + m\angle 3 = 180°$$

Identifying Vertical Angles

EXAMPLE 1

Identify the pairs of vertical angles formed by the intersecting lines to the right.

You can solve this problem using the definition for vertical angles.

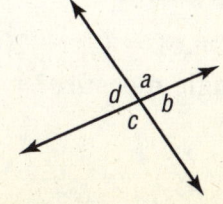

The two pairs of vertical or opposite angles are $\angle a$ and $\angle c$, and $\angle b$ and $\angle d$.

▶ **Understanding the Solution** You can recognize vertical angles as angles that are opposite each other, and not adjacent.

TRY IT!

Identify the vertical angles in the figure.

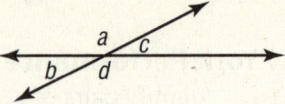

Using Relationships Between Vertical Angles

EXAMPLE 2

If $m\angle f = (2x + 12)°$ and $m\angle c = (3x - 15)°$, what is the value of each angle measure?

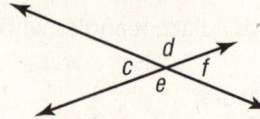

You can solve this problem by applying the Vertical Angles Theorem.

From the Vertical Angles Theorem, you know that if two angles are vertical angles, then they are congruent and their measures are equal. Angles f and c are vertical angles, so you can set the given expressions for their measures equal and solve for x.

$$m\angle f = m\angle c$$
$$2x + 12 = 3x - 15$$
$$12 = x - 15$$
$$27 = x$$

Now substitute 27 for x to determine the value of each angle measure.

$$m\angle f = 2x + 12 = 2(27) + 12 = 66°$$
$$m\angle c = 3x - 15 = 3(27) - 15 = 66°$$

Both angles measure 66°: $m\angle f = m\angle c = 66°$.

▶ **Understanding the Solution** Since the two angles are vertical angles, they should be equal in measure. Check your calculations. Substituting your answer for x should give you the same value for both expressions.

TRY IT!

If $m\angle 1 = (2x + 20)°$ and $m\angle 3 = (x + 60)°$, what is the value of each angle measure?

In the figure, if $m\angle 1 = 45°$, find $m\angle 2$.

Angles 1 and 2 are adjacent angles, so the sum of their measures is 180°.

$$45° + m\angle 2 = 180°$$
$$m\angle 2 = 180° - 45°$$
$$= 135°$$

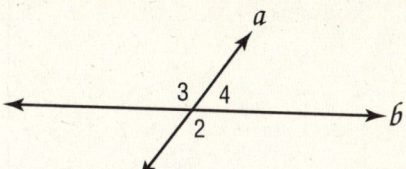

▶ **Understanding the Solution** Angles 1 and 2 are adjacent angles on line b. Together, they form a linear pair of angles with a combined measure of 180°.

Using the figure in Example 3, show that $m\angle 1 = m\angle 4$ without using the Vertical Angles Theorem.

Exercises

SHORT RESPONSE

1 The sum of the measures of a linear pair of angles is equal to 180°. Use this information and the Vertical Angles Theorem to find the measures of all four angles formed below.

$$m\angle a = (3x - 11)°$$
$$m\angle c = (2x + 33)°$$

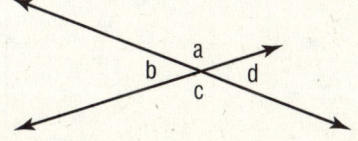

Show your work.

$m\angle a = $ _____ $m\angle b = $ _____ $m\angle c = $ _____ $m\angle d = $ _____

2 Angles 1 and 2 are adjacent angles on a line. What is the sum of the measures of the angles?

A 30°

B 60°

C 90°

D 180°

3 A kite is built by crossing and joining two sticks. If one of the angles formed at the intersection of these sticks measures 73°, what is the measure of the opposite angle?

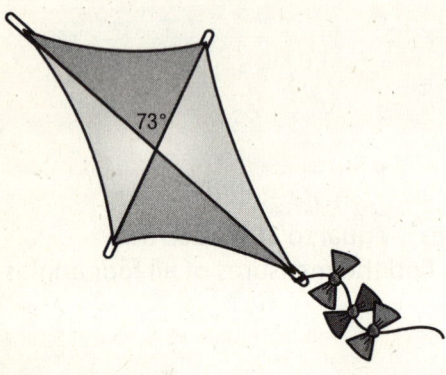

F 17°

G 37°

H 73°

J 107°

4 Using the figure, find x.

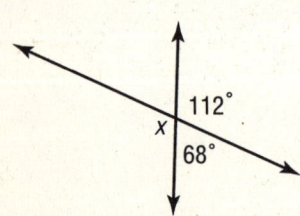

A 22°

B 44°

C 68°

D 112°

5 In the figure, if $m\angle a = 40°$, which of the following is true?

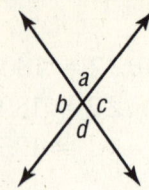

F $m\angle b = 40°$

G $m\angle c = 40°$

H $m\angle d = 40°$

J $m\angle d = 140°$

6 Beth used a pizza cutter to make 2 intersecting cuts and divide a pizza into 4 slices. She expressed the angle of one slice as $(4x + 20)°$ and the one opposite as $(3x + 35)°$.

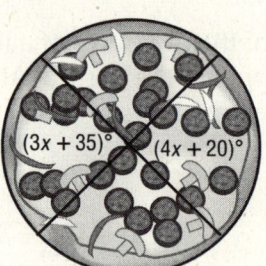

What was the measure of each angle?

A 15°

B 45°

C 60°

D 80°

7 If $m\angle x = (2y + 25)°$, find the value of y.

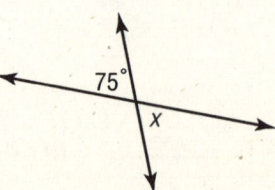

F 21.5

G 25

H 68

J 75

EXTENDED RESPONSE

8 Look at figures A and B below.

Figure A

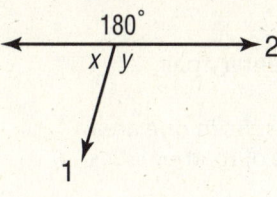

Figure B

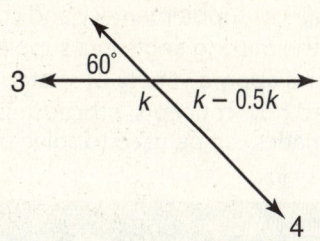

Part A

For which figure(s) can you use what you know about vertical angles to find the measures of the angles? Explain your answer on the lines below.

Part B

Use the Vertical Angles Theorem to find the measures of the angles in the figure(s) you selected above. Sketch the figure(s) in the space below and label each angle with its measurement.

Show your work.

LESSON 6.2 Identifying Supplementary and Complementary Angles

 New York Performance Indicators

8.G.2 Identify pairs of supplementary and complementary angles
8.G.3 Calculate the missing angle in a supplementary or complementary pair
8.CN.3 Connect and apply a variety of strategies to solve problems
8.CN.5 Understand how concepts, procedures, and mathematical results in one area of mathematics can be used to solve problems in other areas of mathematics

VOCABULARY

Complementary angles are two angles whose measures add up to 90°.
Supplementary angles are two angles whose measures total 180°.

REVIEW

Understanding Complementary and Supplementary Angles

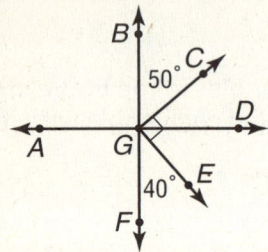

$m\angle BGC + m\angle CGD = 90°$, so $\angle BGC$ and $\angle CGD$ are complementary angles.

$m\angle AGE + m\angle EGD = 180°$, so $\angle AGE$ and $\angle EGD$ are supplementary angles.

What You Should Know

Two angles that form a linear pair are always supplementary to each other.

$\angle AGE$ and $\angle EGD$ form a linear pair and are supplementary along line $\overline{AD}$.

However, complementary and supplementary angles may or may not be adjacent angles.

$\angle AGB$ and $\angle CGE$ are supplementary but not adjacent.

$\angle BGC$ and $\angle FGE$ are complementary but not adjacent.

Using Definitions to Identify Angles

EXAMPLE 1

Lines r and ℓ are perpendicular $r \perp \ell$. Are angle z and the 57° angle complementary, supplementary, or neither? Find the measure of angle z.

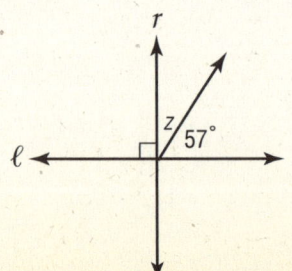

You can solve this problem using the definitions for perpendicular lines and complementary and supplementary angles.

If lines r and ℓ are perpendicular, then they must intersect at right or 90° angles. Therefore, angle z and the 57° angle are complementary and the sum of their measures is 90°.

$$m\angle z + 57 = 90$$
$$m\angle z = 90 - 57$$
$$m\angle z = 33°$$

▶ **Understanding the Solution** Looking at the figure, you can see that lines r and ℓ form four right angles at their intersection. Since angle z and the 57° angle make up one of these right angles, they are complementary.

TRY IT!

What is the complement of an angle measuring 41°?

EXAMPLE 2

The two angles below are supplementary. What is the value of x?

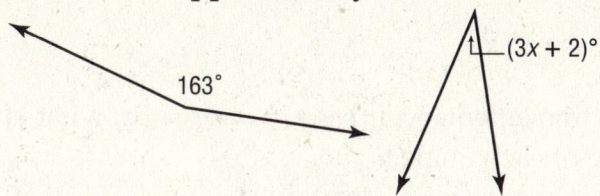

You can solve this problem using the definition of supplementary angles.

$$163 + (3x + 2) = 180$$
$$3x + 2 = 17$$
$$x = 5$$

If you substitute 5 for x in $3x + 2$, you will find that the measure of that angle is 17°.

▶ **Understanding the Solution** You were given that the two angles are supplementary. If the problem had asked you for the measure of the second angle, you could have just subtracted 163° from 180°.

TRY IT!

Angles s and t are supplementary. Angle s measures 29°. Angle t measures $(14x + 11)°$. What is the value of x and $m\angle t$?

EXAMPLE 3

Angles *r* and *s* are supplementary. If *m∠r* = (3*x* − 10)° and *m∠s* = (2*x* + 25)°, what is the value of each angle measure?

You can solve this problem using what you know about supplementary angles.

Set the sum of both expressions equal to 180° and solve for *x*.

$$(3x - 10) + (2x + 25) = 180$$
$$5x + 15 = 180$$
$$5x = 165$$
$$x = 33$$

Now substitute the result for *x* in both expressions to find the measure of each angle.

$$m\angle r = 3x - 10 = 3(33) - 10 = 89°$$
$$m\angle s = 2x + 25 = 2(33) + 25 = 91°$$

▶ **Understanding the Solution** Knowing the definition of supplementary angles is the key to solving the problem. Check your answers by adding the two angle measures you found to verify that their sum is 180°: 89° + 91° = 180°, so angles *r* and *s* are supplementary.

TRY IT!

Angles *y* and *z* are complementary. If *m∠y* = *m∠z* = 5*x*, what is the value of *x* and the measure of each angle?

EXAMPLE 4

What are the measures of ∠*TUV* and ∠*VUW*?

You can solve this problem by determining whether the angles are supplementary.

The angles are adjacent because they share the same vertex, *U*, and a common side, $\overrightarrow{UV}$. Since $\overrightarrow{UT}$ and $\overrightarrow{UW}$ lie on the same line, the two angles are a linear pair and supplementary.

Set the sum of both expressions equal to 180° and solve for *x* to find *m∠VUW*.

$$3x + 12 + x = 180$$
$$4x = 168$$
$$x = 42$$
$$m\angle VUW = 42°$$

Now substitute the value you found for x in the expression for $m\angle TUV$.
$$m\angle TUV = 3(42) + 12 = 126 + 12 = 138°$$

▶ **Understanding the Solution** A linear pair of angles forms a straight angle, and the measure of a straight angle is 180°. Check your answers to make sure that the two angles are supplementary.

TRY IT!

Find the value of x and the measures of the labeled angles in the rectangle below.

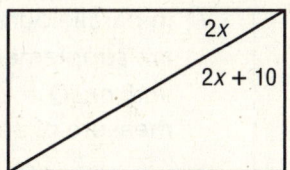

Exercises

SHORT RESPONSE

1 In a trapezoid, two consecutive angles on different bases are supplementary. In the figure below, angles L and M are supplementary.

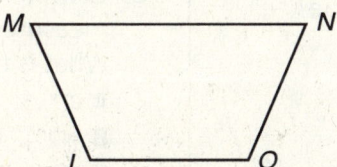

If $m\angle L = (3x + 4)°$ and $m\angle M = (2x - 34)°$, find the measure of each angle.

Show your work.

$m\angle L =$ _____ $m\angle M =$ _____

2 If $m\angle a = 86°$, what is the measure of its complement?

   A 4°

   B 40°

   C 68°

   D 94°

3 What is the measure of angle s on the clock below?

   F 23°

   G 67°

   H 93°

   J 113°

4 Based on the drawing below, which of the following statements is true?

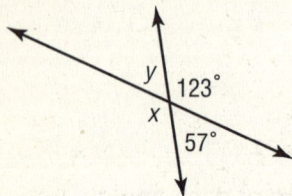

   A $\angle x$ and the 123° angle are supplementary.

   B $\angle y$ and the 123° angle are supplementary.

   C $\angle x$ and the 57° angle are complementary.

   D $\angle y$ and the 57° angle are complementary.

5 If the triangle shown is a right triangle, what is the value of x?

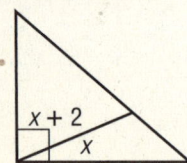

   F 22

   G 44

   H 46

   J 88

6 Suppose $m\angle r = (4x + 11)°$ and $m\angle s = (x + 14)°$. If angles r and s are supplementary, what is the measure of the smaller angle?

   A 31°

   B 42°

   C 45°

   D 135°

7 In parallelogram SRQT, angles R and Q are supplementary. If $m\angle R = (3x + 22)°$ and $m\angle Q = (5x - 2)°$, what is the measure of angle Q?

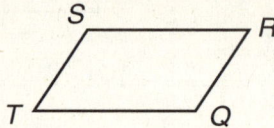

   F 20°

   G 70°

   H 82°

   J 98°

8 Complementary angles m and n have measures of $(3y + 21)°$ and $(4y + 20)°$. What is the value of the larger angle?

   A 7°

   B 42°

   C 48°

   D 49°

9 The measures of two angles are $4x°$ and $(2x + 12)°$. What must be true if x = 13?

   F Both angles are acute and complementary.

   G Both angles are obtuse and supplementary.

   H One angle is acute, one is obtuse, and they are complementary.

   J One angle is acute, one is obtuse, and they are supplementary.

10 In the figure below, $\ell \perp m$. Use two methods to find the measures of the angles *f* and *e*.

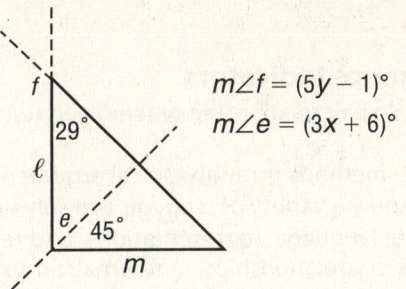

$m\angle f = (5y - 1)°$

$m\angle e = (3x + 6)°$

Part A

Find the measure of each angle using angle definitions. In your calculations, list the definitions you used.

Show your work.

m∠e = _____   *m∠f =* _____

Part B

Now use the same definitions to write and solve equations for *x* and *y*. Substitute the values for the variables into the expressions to find the measures of the angles.

Show your work.

m∠e = (3x + 6)° = _____   *m∠f = (5y − 1)° =* _____

Relating Angles Formed by a Transversal and Two Parallel Lines

New York Performance Indicators

8.G.4 Determine angle pair relationship when given two parallel lines cut by a transversal

8.PS.16 Justify solution methods through logical argument

8.CN.3 Connect and apply a variety of strategies to solve problems

8.CM.10 Use appropriate language, representations, and terminology when describing objects, relationships, mathematical solutions, and rationale

VOCABULARY

Parallel lines are lines in the same plane that never intersect.

A **transversal** is a line that intersects two or more lines.

The angles formed by parallel lines and a transversal are: **Interior angles, Exterior angles, Alternate Interior angles, Alternate Exterior angles,** and **Corresponding angles.**

REVIEW

Understanding Angle Relationships

Lines ℓ and m are parallel, or $\ell \parallel m$. Line s is the transversal.

A transversal forms eight angles with the two parallel lines, as shown.

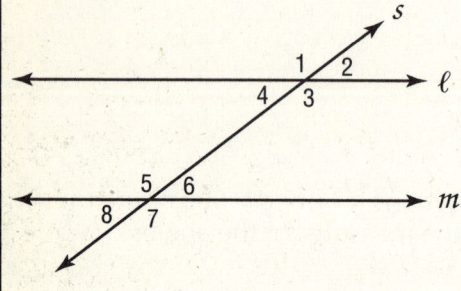

What You Should Know

A transversal creates pairs of congruent angles.

Interior angles are formed between the parallel lines. In the figure to the left, they are $\angle 3$, $\angle 4$, $\angle 5$, and $\angle 6$.

Exterior angles are formed on the outer sides of the parallel lines. They are $\angle 1$, $\angle 2$, $\angle 7$, and $\angle 8$.

Corresponding angles are pairs of nonadjacent angles on the same side of the transversal. They are $\angle 1$ and $\angle 5$, $\angle 4$ and $\angle 8$, $\angle 2$ and $\angle 6$, and $\angle 3$ and $\angle 7$.

Identifying Corresponding Angles

When a transversal cuts two lines that are parallel, all four pairs of corresponding angles are congruent. In the figure below, notice that the sides of a pair of corresponding angles form an F-shape.

Corresponding Angles

$\angle 1 \cong \angle 5$

$\angle 2 \cong \angle 6$

$\angle 3 \cong \angle 7$

$\angle 4 \cong \angle 8$

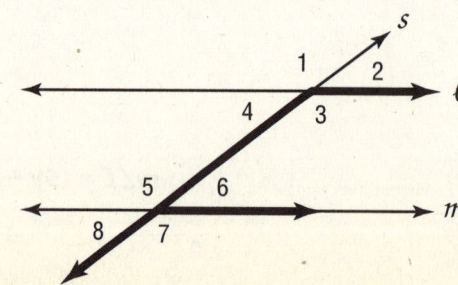

EXAMPLE 1

Name all sets of corresponding angles in the figure below, where $\ell \parallel m$. If $m\angle d = 70°$, what is $m\angle h$?

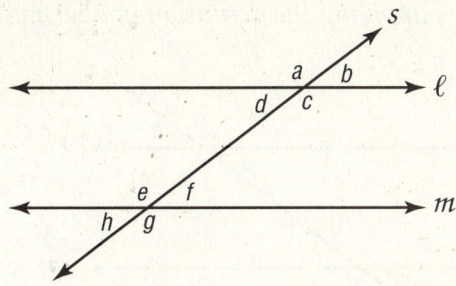

You can solve this problem by applying the definition of corresponding angles.

To help you identify pairs of corresponding angles, look for angles that form F-shapes with their sides.

The corresponding angles are $\angle a$ and $\angle e$, $\angle b$ and $\angle f$, $\angle c$ and $\angle g$, $\angle d$ and $\angle h$.

Since $\angle d$ and $\angle h$ are corresponding angles, they are also congruent.

$$\angle d \cong \angle h, \text{ so } m\angle d = m\angle h$$
$$m\angle d = 70°, \text{ so } m\angle h = 70°$$

▶ **Understanding the Solution** The sides of corresponding angles can form a normal letter F or a backward F. Draw the F backward too so that you get all the pairs. Remember that congruent angles have equal measure.

TRY IT!

Using the same figure in Example 1, if $m\angle b = 70°$, what is $m\angle f$?

Identifying Alternate Interior and Alternate Exterior Angles

Alternate interior angles are a pair of nonadjacent interior angles on opposite sides of the transversal. Notice that a Z-shape can be used to recognize alternate interior angles. Alternate exterior angles are on opposite sides of the transversal and outside the parallel lines.

When a transversal cuts two lines that are parallel, the alternate interior angles are congruent. The alternate exterior angles are also congruent.

Alternate Interior Angles

$\angle 3 \cong \angle 5$

$\angle 4 \cong \angle 6$

Alternate Exterior Angles

$\angle 1 \cong \angle 7$

$\angle 2 \cong \angle 8$

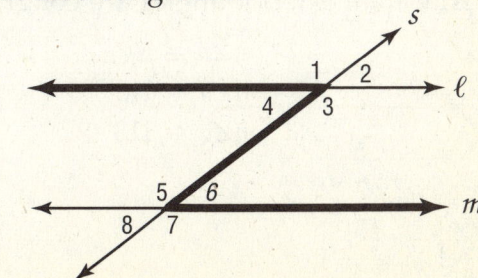

EXAMPLE 2

Identify the alternate interior angles in the figure below, where $\ell \parallel m$. If $m\angle c = 118°$, what is $m\angle e$?

You can solve this problem by applying the definition of alternate interior angles.

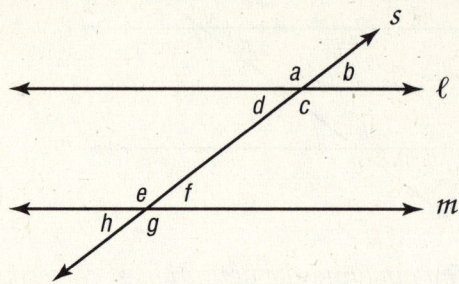

Look for angles that form Z-shapes with their sides.

The alternate interior angles are $\angle c$ and $\angle e$, and $\angle d$ and $\angle f$.

Since $\angle c$ and $\angle e$ are alternate interior angles, they are congruent.

$$\angle c \cong \angle e$$
$$m\angle c = 118°$$
$$m\angle e = 118°$$

▶ **Understanding the Solution** The key to this problem is recognizing the alternate interior angles and remembering that the angles are congruent. You can draw the letter Z to find the angles.

TRY IT!

Using the same figure in Example 2, if $m\angle f = 62°$, what is $m\angle d$?

EXAMPLE 3

In the figure below, $\ell \parallel m$ and $m\angle a = 113°$. What is $m\angle g$?

You can solve this problem by identifying a relationship between the two angles.

Angles a and g are on opposite sides of the transversal. Since they also lie outside the parallel lines, they are alternate exterior angles.

Alternate exterior angles are congruent.

$$\angle a \cong \angle g$$
$$m\angle a = 113°$$
$$m\angle g = 113°$$

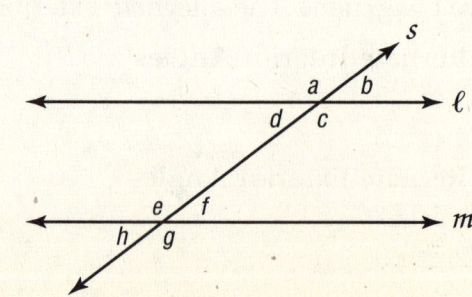

▶ **Understanding the Solution** Check your result by identifying the shape of the two angles. Both angles are obtuse, or greater than 90°.

TRY IT!

Using the same figure in Example 3, if $m\angle h = 67°$, what is $m\angle b$?

Exercises

SHORT RESPONSE

In the figure below, $\ell \parallel m$. Use the figure to answer questions 1 and 2.

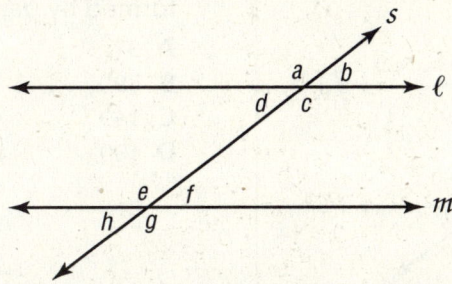

1 If $m\angle a = 123°$, find the measures of $\angle b$, $\angle e$, and $\angle g$. State the relationship of each angle to $\angle a$ on the lines below.

$m\angle b =$ _____ $m\angle e =$ _____ $m\angle g =$ _____

2 If $m\angle d = 57°$, find the measures of $\angle b$, $\angle f$, $\angle a$, and $\angle h$. State the relationship of each angle to $\angle d$ on the lines below.

$m\angle b =$ _____ $m\angle f =$ _____

$m\angle a =$ _____ $m\angle h =$ _____

3 Angles *x* and *y* are alternate interior angles formed by two parallel lines and a transversal. If $m\angle x = 167°$, what is $m\angle y$?

A 13°
B 77°
C 167°
D 180°

In the figure below, j ∥ k. Use the figure to answer questions 4 and 5.

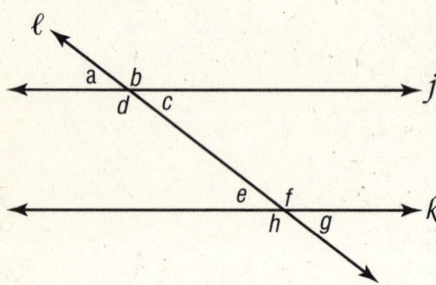

4 Which of the following angles are congruent to ∠*d*?

F ∠*c*, ∠*e*, ∠*g*
G ∠*b*, ∠*h*, ∠*f*
H ∠*a*, ∠*e*, ∠*h*
J ∠*h* only

5 If $m\angle c = 64°$, what is the measure of ∠*e* and how are the two angles related?

A 64°, alternate interior angles
B 64°, alternate exterior angles
C 116°, alternate interior angles
D 116°, supplementary angles

6 Which of the following statements is true when two parallel lines are crossed by a transversal?

F All interior angles are congruent.
G All exterior angles are congruent.
H Same-side interior angles are congruent.
J Alternate interior angles are congruent.

7 Angle *t* measures 149°. What is the measure of corresponding angle *s* formed by parallels and a transversal?

A 31°
B 59°
C 149°
D 180°

8 On the graph below, the line $y = 3$ is parallel to the *x*-axis. If the line $y = x$ is a transversal, what are the measures of the interior angles?

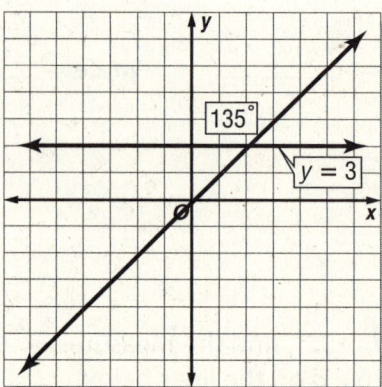

F 45° and 135°
G 55° and 125°
H 70° and 110°
J 85° and 95°

9 In the figure, lines ℓ and m are parallel.

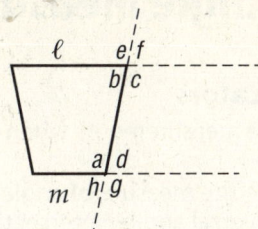

Part A

Name all the angle-pair relationships in the figure and list the angle pairs for each.

_____

_____

_____

_____

Part B

In the figure, $\angle f \cong \angle h$. Explain why the two angles are congruent using two of the angle relationships you listed above.

Applying Algebra to Find Angle Measures

New York Performance Indicators

8.G.5 Calculate the missing angle measurements when given two parallel lines cut by a transversal

8.A.12 Apply algebra to determine the measure of angles formed by or contained in parallel lines cut by a transversal and by intersecting lines

8.R.6 Use representations to explore problem situations

8.CN.5 Understand how concepts, procedures, and mathematical results in one area of mathematics can be used to solve problems in other areas of mathematics

---

REVIEW

Understanding Parallel Lines Cut by a Transversal

Besides forming congruent angles, a transversal intersecting two parallel lines also forms several pairs of nonadjacent supplementary angles.

Applying Algebra

When an unknown angle is represented by an algebraic expression, you can use the relationships between angles to find the measure of a missing angle.

Using the Sum of Interior Angles

In the figure below, $\ell \parallel m$. On each side of the transversal, the interior angles form a pair of supplementary angles. The sum of the measures of the two interior angles on each side of the transversal is 180°.

$\angle 4$ and $\angle 5$ are interior angles on the left side of the transversal.

$$m\angle 4 + m\angle 5 = 180°$$

$\angle 3$ and $\angle 6$ are interior angles on the right side of the transversal.

$$m\angle 3 + m\angle 6 = 180°$$

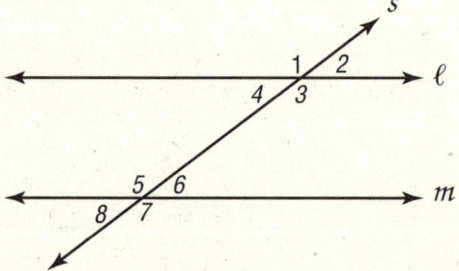

EXAMPLE 1

In the figure below, $\ell \parallel m$. Find the measure of $\angle f$ that is given by the expression $2x - 10$.

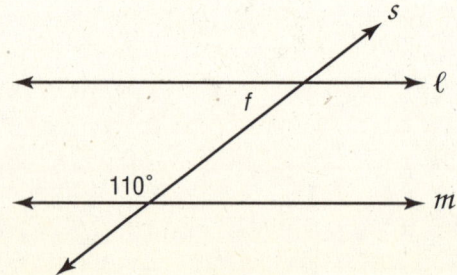

You can solve this problem by identifying congruent or supplementary angles.

The angles with measures $(2x - 10)°$ and $110°$ are interior angles that lie on the same side of the transversal. So they are supplementary angles.

$$(2x - 10) + 110 = 180$$
$$2x = 180 - 100$$
$$2x = 80$$
$$x = 40$$

Now substitute the value you found for x in the expression for $m\angle f$.

$$m\angle f = 2x - 10 = 2(40) - 10 = 70°$$

▶ **Understanding the Solution** Check your result: $110° + 70° = 180°$.

TRY IT!

In the figure below, line ℓ is parallel to line m. Find the angle measure that is given by the expression $3x$.

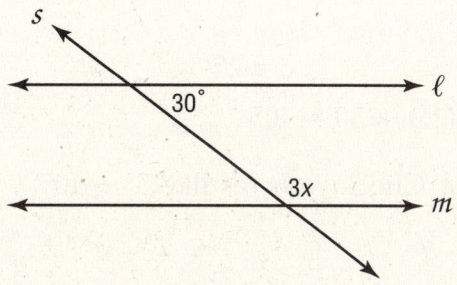

Using the Sum of Exterior Angles

On each side of the transversal, the exterior angles form a pair of supplementary

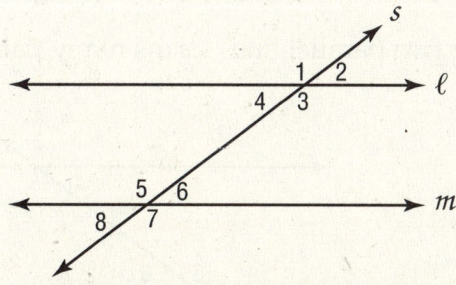

angles. The sum of the measures of each pair of exterior angles is $180°$.

In the figure above, $\ell \parallel m$.

$\angle 1$ and $\angle 8$ are exterior angles on the left side of the transversal.

$$m\angle 1 + m\angle 8 = 180°$$

$\angle 2$ and $\angle 7$ are exterior angles on the right side of the transversal.

$$m\angle 2 + m\angle 7 = 180°$$

EXAMPLE 2

In the figure below, $m\angle a = (3x + 30)°$ and $m\angle h = 75°$. Find the measure of $\angle a$.

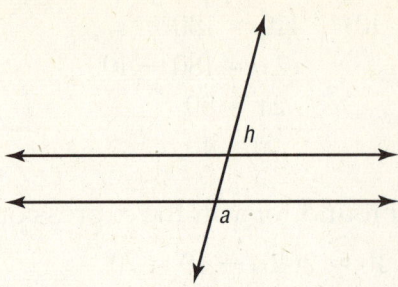

You can solve this problem by identifying congruent or supplementary angles.

Angles a and h are exterior angles that lie on the same side of the transversal. Since they are supplementary angles, you can set the sum of their measures equal to 180 degrees and solve for x.

$$(3x + 30) + 75 = 180$$
$$3x = 180 - 105$$
$$3x = 75$$
$$x = 25$$

Find $m\angle a$: $3x + 30 = 3(25) + 30 = 105°$.

▶ **Understanding the Solution** Check your result: $75° + 105° = 180°$.

TRY IT!

Using the figure from Example 2, find the measure of the angle corresponding to $\angle a$.

Using Congruent Angles

A transversal intersecting two parallel lines forms many pairs of congruent angles.

Corresponding Angles

$\angle 1 \cong \angle 5$
$\angle 2 \cong \angle 6$
$\angle 3 \cong \angle 7$
$\angle 4 \cong \angle 8$

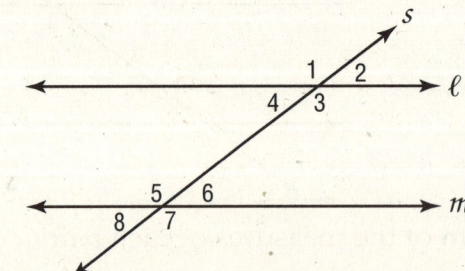

Alternate Interior Angles

$\angle 3 \cong \angle 5$
$\angle 4 \cong \angle 6$

Alternate Exterior Angles

$\angle 1 \cong \angle 7$
$\angle 2 \cong \angle 8$

In the figure below, $\ell \parallel m$, $m\angle 4 = (x + 10)°$ and $m\angle 8 = (2x - 30)°$. Find the measure of angle 6.

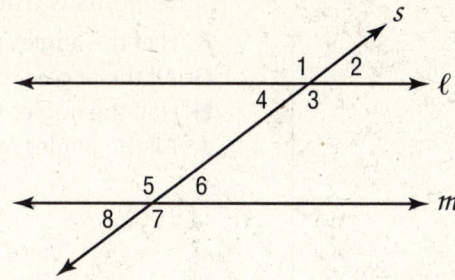

You can solve this problem by identifying congruent angles.

$\angle 8$ and $\angle 4$ are corresponding angles, so they are congruent.

$$2x - 30 = x + 10$$
$$2x - x = 30 + 10$$
$$x = 40$$

Using substitution, $m\angle 4 = x + 10 = 50°$.

Since $\angle 4$ and $\angle 6$ are alternate interior angles, they are congruent. So $m\angle 6 = 50°$.

▶ **Understanding the Solution** Angles 8 and 6 are vertical angles, so they are also congruent. Check your result by substituting 40 for x in the expression for $m\angle 8$:

$$m\angle 8 = 2x - 30 = 2(40) - 30 = 50, \text{ so } m\angle 6 = 50°.$$

Find $m\angle 2$ in the figure for Example 3.

Exercises

SHORT RESPONSE

1 Suppose lines ℓ and m are parallel, and $m\angle b = 45°$. Identify all angles that are complementary to $\angle b$. Explain your answer on the lines below.

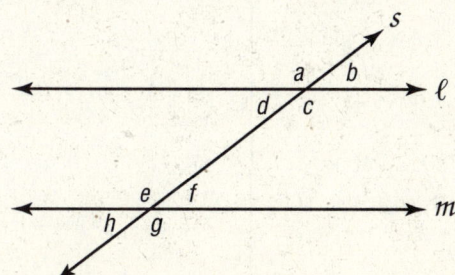

2 How many pairs of corresponding angles are formed when a transversal intersects two lines?

  A 2
  B 3
  C 4
  D 6

In the figure below, $j \parallel k$. Use the figure to answer questions 3 and 4.

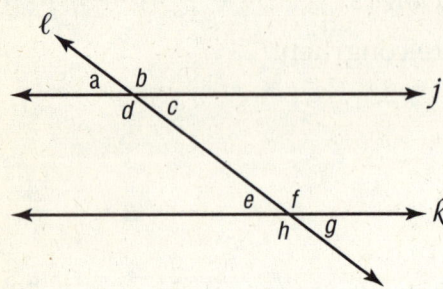

$$m\angle d = 2x°$$
$$m\angle h = (x + 50)°$$

3 Which of the following equations describes the relationship between $\angle d$ and $\angle h$?

  F $2x + (x + 50) = 180°$
  G $2x + (x + 50) = 90°$
  H $2x = x + 50$
  J $2x - (x + 50) = 180°$

4 What is the measure of $\angle h$?

  A 120°
  B 100°
  C 82°
  D 50°

5 A transversal intersects two parallel lines. The measure of one of the angles formed is 90°. Which of the following statements is true?

  F Half the angles formed will be acute.
  G All the angles formed are acute.
  H Half the angles formed are obtuse.
  J All the angles formed are right angles.

6 Angles $\angle j$ and $\angle k$ are vertical angles. Angle $\angle p$ is supplementary to $\angle k$. Which of the following statements is **not** true?

  A $m\angle j = m\angle k$
  B $m\angle p + m\angle k = 180°$
  C $m\angle p + m\angle k = 90°$
  D $m\angle p + m\angle j = 180°$

7 Two parallel lines are cut by a transversal. All of the following angles are congruent, **except**

  F corresponding angles.
  G alternate interior angles.
  H same-side exterior angles.
  J alternate exterior angles.

8 In the figure below, $\ell \parallel m$. Use the figure to answer Parts A and B below.

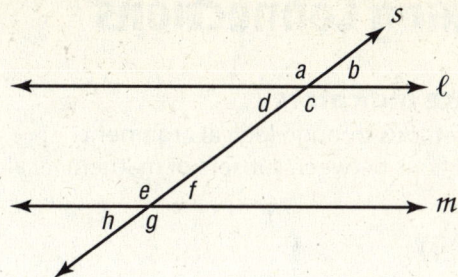

Part A

In the figure, $m\angle d = (x + 30)°$ and $m\angle f = (2x - 5)°$. Find the measures of angles d and f.

Show your work.

$m\angle d =$ _____ $m\angle f =$ _____

Part B

Identify all the angles that are congruent to $\angle c$, and explain why they are congruent.

LESSON 6.5 *Problem-Solving Strategy:*
Making Connections

New York Performance Indicators

8.PS.16 Justify solution methods through logical argument
8.CN.2 Recognize connections between subsets of mathematical ideas

Understand the Strategy

You can apply the properties of parallel lines cut by a transversal to find
the interior angles in quadrilaterals with at least one pair of parallel sides.

Problem: Find the measures of angles *B* and *D* in the trapezoid.

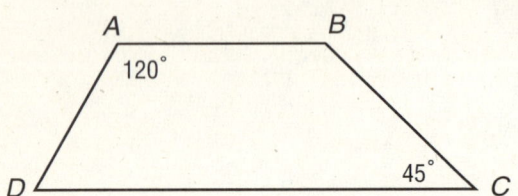

SOLUTION

What do you know?

$\overline{AB}$ and $\overline{DC}$ are parallel.
$m\angle A = 120°$
$m\angle C = 45°$

What do you need to find?

$m\angle B$ and $m\angle D$

Find the relationship.

Since $\overline{AB}$ and $\overline{DC}$ are parallel, the other two sides are transversals. Pick
one of these sides to work with, such as $\overline{AD}$. Look at the figure and relate
its angles to the interior angles formed by a transversal that intersects two
parallel lines.

Interior angles *A* and *D* are ⟶ $m\angle A + m\angle D = 180$
supplementary, or add up to 180°.　$120 + m\angle D = 180$
$m\angle D = 180 - 120 = 60°$

Interior angles *B* and *C* are ⟶ $m\angle C + m\angle B = 180$
also supplementary.　$45 + m\angle B = 180$
$m\angle B = 180 - 45 = 135°$

Check your solution. A trapezoid is a polygon with 4 sides, or a
quadrilateral. The sum of the interior angles in a quadrilateral is 360°.

$$m\angle A + m\angle B + m\angle C + m\angle D = 120° + 135° + 45° + 60° = 360°$$

▶ **Understanding the Solution**　A trapezoid has two pairs of supplementary
angles.

1 Find the measure of ∠*K* in the parallelogram *JMLK*.

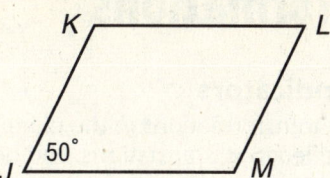

Show your work.

m∠*K* = _____

2 Triangle *ABC* is an equilateral triangle. *DEBA* is a parallelogram. Find the measures of ∠*EBA* and ∠*DAB* in the parallelogram.

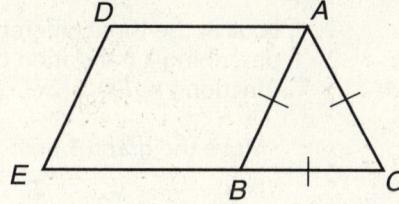

Show your work.

m∠*EBA* = _____

m∠*DAB* = _____

LESSON 6.6   Identifying and Describing Transformations

New York Performance Indicators

8.G.7 Describe and identify transformations in the plane, using proper function notation (rotations, reflections, translations, and dilations)

8.PS.3 Understand and demonstrate how written symbols represent mathematical ideas

8.CM.10 Use appropriate language, representations, and terminology when describing objects, relationships, mathematical solutions, and rationale

VOCABULARY

A **transformation** is a change in the location, size, or orientation of a figure.

Rotation is turning a figure around a fixed point.

Reflection is flipping a figure or finding a mirror image of a figure over a line or over a point.

Translation is sliding a figure to a different location.

Dilation is enlarging or reducing the size of a figure without changing its form or shape.

REVIEW

Understanding Transformations

Transformations can be applied to shapes, geometric figures, or graphs of a function.

In a coordinate plane, the ordered pair (x, y) represents a point on the graph of a function or a geometric figure. You can also use an ordered pair of numbers to describe a transformation.

What You Should Know

An image is the figure that results from a transformation.

Look at the three different ways of describing a translation of a linear function $y = f(x) = 2x$.

• Move the graph 3 units up.

• $(x, y) \longrightarrow (x, y + 3)$

• $y = f(x) + 3$

Identifying Transformations

Compare each original figure with its image.

In a **translation,** a figure slides up or down, or left or right.

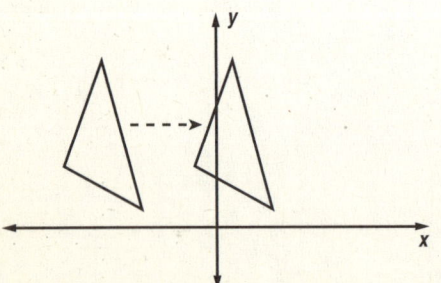

No change in shape or size. The location changes.

In a **reflection,** a mirror image of the figure is formed across a line called the **line of symmetry**.

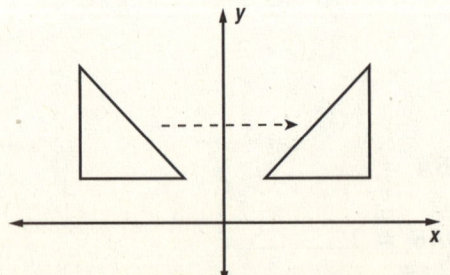

No change in size. The orientation of the shape changes.

In a **rotation,** a figure turns around a fixed point, such as the origin.

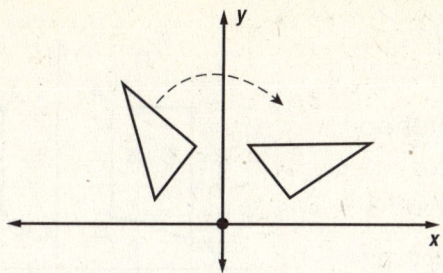

No change in shape, but the orientation and location change.

In a **dilation**, a figure is enlarged or reduced proportionally.

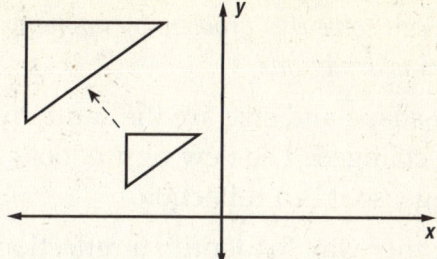

No change in shape, but unlike other transformations, **the size changes.**

EXAMPLE 1

Describe the transformation of the function
$f(x) = 2x$ shown to the right.

You can solve this problem by observing how the coordinates change.

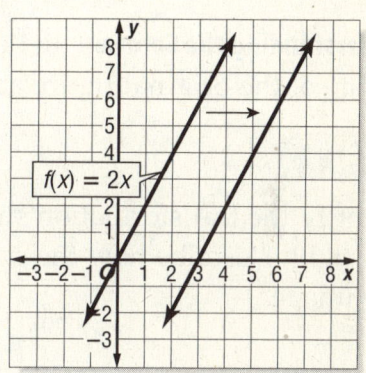

Old Coordinates	New Coordinates
(0, 0)	(3, 0)
(1, 2)	(4, 2)
(2, 4)	(5, 4)
(3, 6)	(6, 6)

Sliding a figure results in a translation. When you slide a figure to the right, the *x*-coordinates increase but the *y*-coordinates do not change. This translation of $f(x) = 2x$ moves it 3 units to the right.

You can describe the new function by substituting $x - 3$ for x:

$$f(x) = 2(x - 3) = 2x - 6$$

This equation gives the coordinates of the transformed function.

▶ **Understanding the Solution** When you translate the function up (or down), you simply add (or subtract) a number to the function value. But when you translate left or right *a* units, you have to replace *x* with $x + a$ or $x - a$ in the function.

TRY IT!

Identify the transformation shown in the figure to the right. Write the new function for the line.

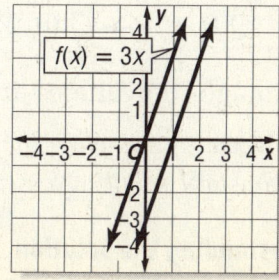

EXAMPLE 2

The figure ABCD is transformed to A'B'C'D'. Identify this transformation.

You can solve this problem by applying the properties of transformations.

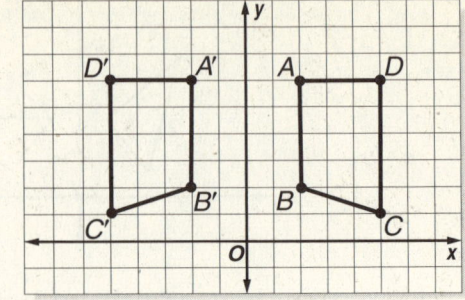

The shape and size are the same, but the orientation has changed. The new figure looks like a mirror image, so it is a reflection.

Another way to identify a reflection is by observing how the coordinates change.

$A(2, 6) \longrightarrow A'(-2, 6)$ $B(2, 2) \longrightarrow B'(-2, 2)$
$C(5, 1) \longrightarrow C'(-5, 1)$ $D(5, 6) \longrightarrow D'(-5, 6)$

To reflect across the y-axis, change the sign of the x-coordinates.

▶ **Understanding the Solution** Reflections are easy to identify. Look for a mirror image of the original figure across a line, such as the x-axis or y-axis.

TRY IT!

Identify the transformation shown in the figure to the right. Describe how the coordinates change in the image.

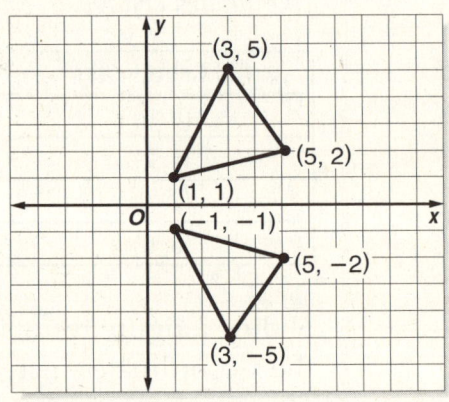

EXAMPLE 3

The figure PQRS is transformed to P'Q'R'S'. Identify and describe this transformation.

The shape of the image is the same but its size is reduced. The image is a dilation about the origin.

Look at how the coordinates change:

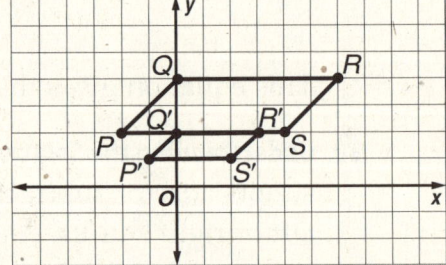

$P(-2, 2) \longrightarrow P'(-1, 1)$ $R(6, 4) \longrightarrow R'(3, 2)$
$Q(0, 4) \longrightarrow Q'(0, 2)$ $S(4, 2) \longrightarrow S'(2, 1)$

The image coordinates are half of the original coordinates.

To reduce a figure using the origin as the center of dilation, divide both the coordinates by a constant number, or scale factor.

▶ **Understanding the Solution** In dilations, all coordinates are divided (or multiplied) by the same number to find the coordinates of the image.

EXAMPLE 4

**Triangle *ABC* is transformed to *A'B'C'*.
Identify and describe this transformation.**

The transformation is not a mirror image or a dilation. In a translation, all points move by the same amount, so this is not a translation either.

The triangle has been rotated clockwise by 90°.

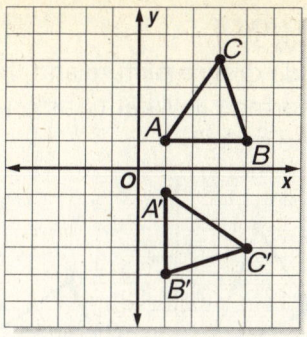

$$A(1, 1) \longrightarrow A'(1, -1) \qquad B(4, 1) \longrightarrow B'(1, -4) \qquad C(3, 4) \longrightarrow C'(4, -3)$$

To rotate a figure clockwise by 90°, switch the coordinates of each point and change the sign of the y-coordinate of the image.

Rules for Rotations About the Origin

Rotation of 90° Clockwise	Rotation of 90° Counterclockwise	Rotation of 180°
Switch the coordinates of each point. Then change the sign of the y-coordinate.	Switch the coordinates of each point. Then change the sign of the x-coordinate	Change the sign of both the x-coordinate and the y-coordinate.

Exercises

SHORT RESPONSE

1 Identify all possible combinations of transformations that could have been made to triangle 1 to get triangle 2. Describe the specific transformations that were made.

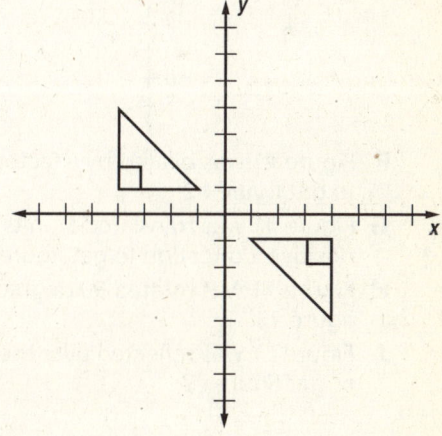

2 Melinda drew a picture and then transformed it. What transformation did she make?

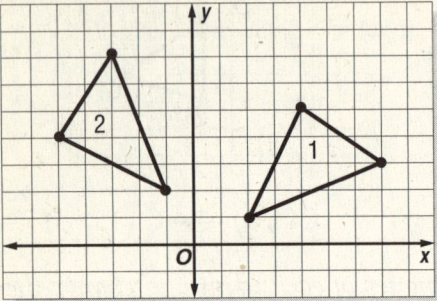

A rotation

B dilation

C translation

D reflection

3 Which of the following is true regarding the figures below?

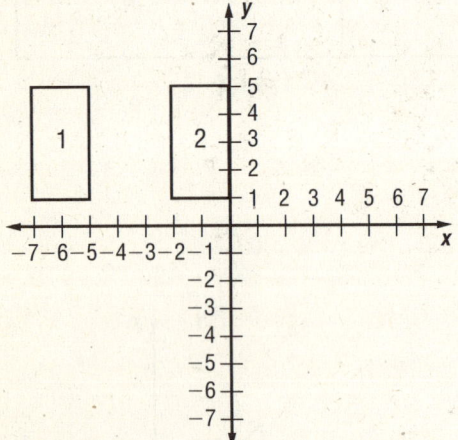

F Figure #1 was dilated by a factor of 3 to get figure #2.

G Figure #1 was translated 5 units in the positive *x* direction to get figure #2.

H Figure #1 was rotated 90 degrees to get figure #2.

J Figure #1 was reflected over the *x*-axis to get figure #2.

4 The coordinates of a triangle are multiplied by 3. Which type of transformation is this?

A rotation

B dilation

C translation

D reflection

5 Kimberly found a star and traced it onto a piece of graph paper. Then she transformed it into figure 2 below.

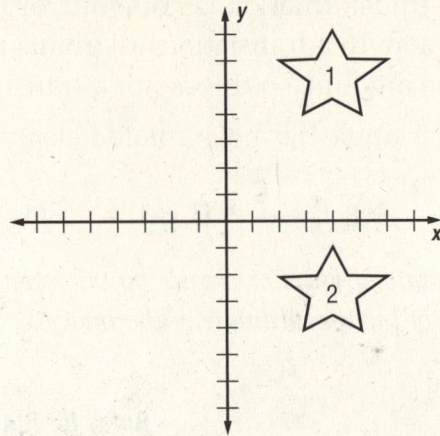

Which transformation did she make?

F dilation

G translation

H rotation

J reflection

6 Which of the following was done to figure 1 to get figure 2?

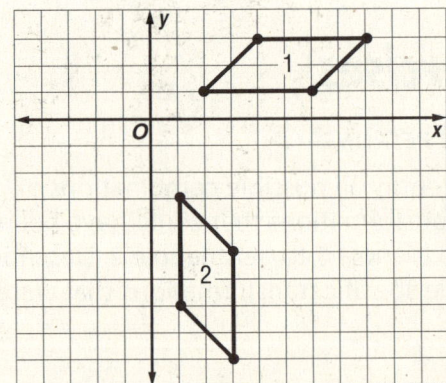

A rotation and translation

B dilation

C reflection and dilation

D rotation

7 The function $f(x) = \frac{1}{2}x$ is translated 3 units up. Which function describes the graph of the image?

F $g(x) = 3\left(\frac{1}{2}x\right)$

G $g(x) = \frac{1}{2}x - 3$

H $g(x) = \frac{1}{2}(x - 3)$

J $g(x) = \frac{1}{2}x + 3$

EXTENDED RESPONSE

8 The function $f(x) = \frac{1}{2}x + 3$ describes a linear function.

Part A

On the lines below, describe and explain the transformation of the function that is represented by $g(x) = \frac{1}{2}x + 1$.

___

___

___

Part B

The function $f(x - 2) = \frac{1}{2}(x - 2) + 3$ describes the transformation of $f(x) = \frac{1}{2}x + 3$ two units to the right. What kind of transformation does $f(x + 1) = \frac{1}{2}(x + 1) + 3$ describe?

___

___

LESSON 6.7 Drawing an Image Under Rotation

New York Performance Indicators

8.G.8 Draw the image of a figure under rotations of 90 and 180 degrees

8.CM.4 Share organized mathematical ideas through the manipulation of objects, numerical tables, drawings, pictures, charts, graphs, tables, diagrams, models and symbols in written and verbal form

8.R.1 Use physical objects, drawings, charts, tables, graphs, symbols, equations, or objects created using technology as representations

8.R.11 Use mathematics to show and understand mathematical phenomena (e.g., use tables, graphs, and equations to show a pattern underlying a function)

VOCABULARY

Rotation is turning a figure about a fixed point.

REVIEW

Understand Rotations

A figure can be rotated between 0 and 360 degrees around a point.

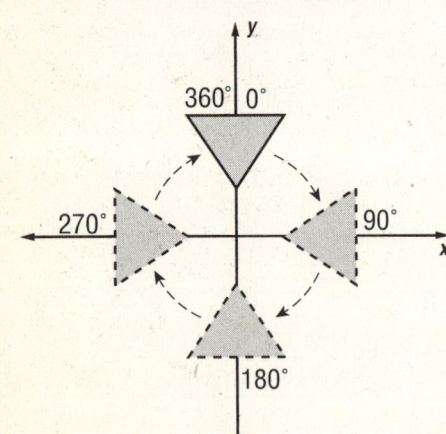

What You Should Know

A figure can be turned in one of two directions:

A **clockwise** turn follows the same direction as the hands of a clock.

A **counterclockwise** turn follows the opposite direction of the hands of a clock.

In the figure to the left, notice that a 360° rotation about the origin brings the figure back to its original position.

Drawing Rotations by Tracing

Rotating a figure 360 degrees in either direction results in an image that is identical to the original. A 360° rotation is also called a **full turn**. Rotating a figure 180° clockwise or counterclockwise is called a **half turn**.

EXAMPLE 1

Draw the image that is formed by rotating △*ABC* clockwise about the origin by 180°.

You can rotate the figure by tracing it and using what you know about 180° rotations.

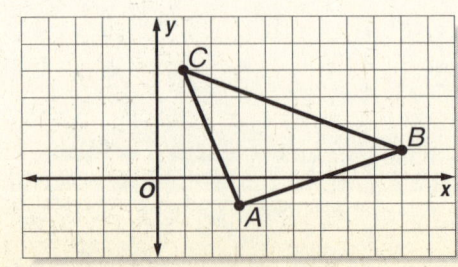

Place a thin piece of paper over △*ABC* and trace the vertices *A*, *B*, and *C*, as well as the *x*- and *y*-axes. Label each vertex and axis.

Press the tip of your pencil on the point (0, 0) to hold the paper in place. Then turn the paper to the right 180°, so that the *x*-axes line up and the *y*-axes line up.

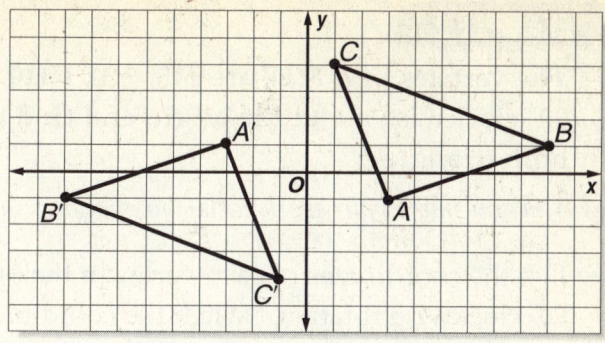

Press hard on the vertices with your pencil to mark the positions of *A'*, *B'*, and *C'*. Now remove the tracing paper and draw line segments to connect the new vertices and draw the image, △*A'B'C'*.

▶ **Understanding the Solution** Since a 360° rotation brings the figure back to its original position, a 180° rotation forms an image that is only halfway around the origin.

TRY IT!

Draw the image that would result from rotating △*DEF* counterclockwise about the origin by 180°.

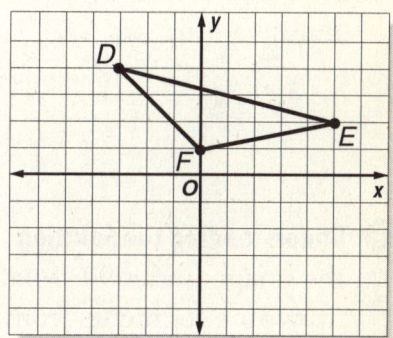

Drawing Rotations Using Ordered Pairs

In addition to tracing, you can also rotate figures 90° or 180° in either direction about the origin using the following rules. Apply the rule to each vertex or given point of the original figure.

<u>Rotation</u>	<u>Rule</u>	<u>Example</u>
90° clockwise	Switch the *x*- and *y*-coordinates. Multiply the new *y*-coordinate by −1.	$(3, 4) \rightarrow (4, -3)$
90° counterclockwise	Switch the *x*- and *y*-coordinates. Multiply the new *x*-coordinate by −1.	$(3, 4) \rightarrow (-4, 3)$
180° in either direction	Multiply both the *x*- and *y*-coordinates by −1.	$(3, 4) \rightarrow (-3, -4)$

EXAMPLE 2

The vertices of △*TUV* are *T*(5, −3), *U*(10, 1), and *V*(8, −7). Rotate the triangle 90° clockwise about the origin and find the coordinates of △*T′U′V′*. Graph both triangles.

You can make your work easier by applying the rule for rotating a figure 90° clockwise.

List the coordinates of each vertex in the original triangle and apply the rule for a 90° clockwise rotation. Switch the coordinates in each pair, and multiply the second or *y*-coordinate by −1. Use the coordinates to graph both △*TUV* and △*T′U′V′*.

$T(5, -3) \longrightarrow T'(-3, -5)$
$U(10, 1) \longrightarrow U'(1, -10)$
$V(8, -7) \longrightarrow V'(-7, -8)$

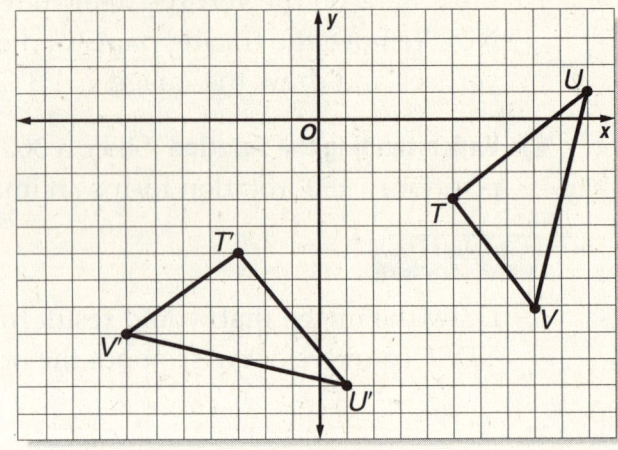

▶ **Understanding the Solution** To check that the image under 90°-rotation is correct, draw line segments from the origin to two corresponding vertices of the figures. For example, draw segments $\overline{RT}$ and $\overline{RT'}$. The angle formed by the two segments should measure 90°, or ∠*TRT′* should be a right angle. Using the other vertices, ∠*VRV′* and ∠*URU′* should also be right angles.

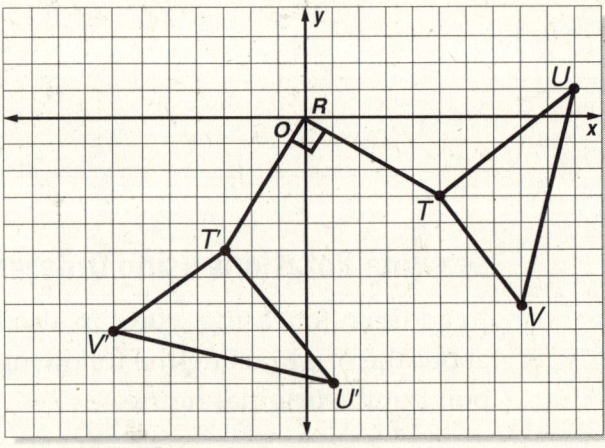

TRY IT!

The vertices of trapezoid *MATH* are *M*(4, 6), *A*(9, 6), *T*(11, 3), and *H*(2, 3). Rotate the figure 90° counterclockwise about the origin and find the coordinates of trapezoid *M′A′T′H′*. Graph both figures on a piece of graph paper.

Exercises

MULTIPLE CHOICE

1 Which figure shows △BCD rotated 180° clockwise about the origin?

A

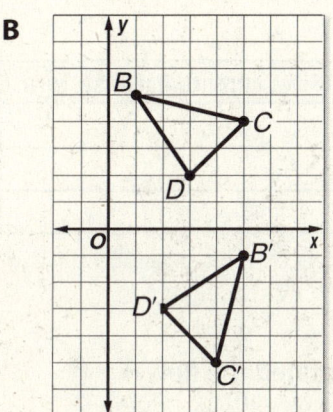

B

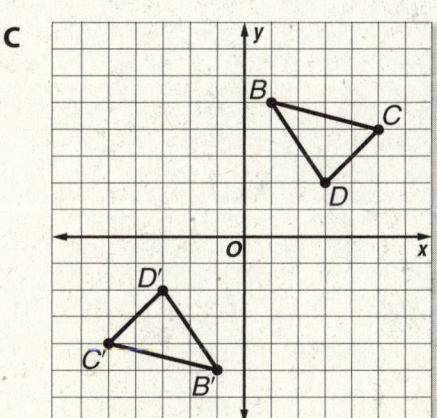

C

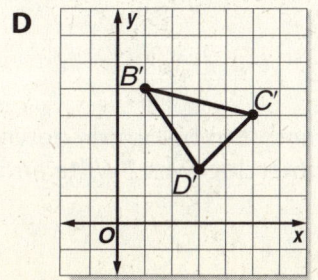

D

2 Figure QRST has vertices Q(2, −1), R(6, −1), S(6, −3), and T(2, −3). Which will be the coordinates of vertex S after the rectangle is rotated 90° counterclockwise about the origin?

F (−6, −3)

G (6, 3)

H (−3, 6)

J (3, 6)

3 Which type of rotation about the origin describes the relationship between figures 1 and 2 below?

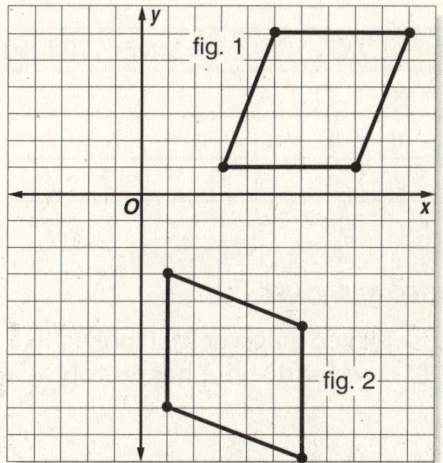

A 90 degrees clockwise

B 90 degrees counterclockwise

C 180 degrees clockwise

D 180 degrees counterclockwise

4 Rotate the quadrilateral shown below 180° about the origin. Which of the following ordered pairs does not belong to any vertex in the rotated image?

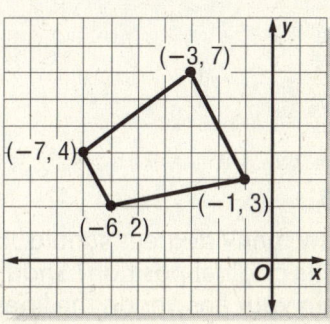

F (7, −4)

G (3, 7)

H (6, −2)

J (1, −3)

5 Graph the 90°-clockwise and 180°-counterclockwise rotations of △HIJ on the coordinate grid below. Write the coordinates of the vertices for each image in the chart below.

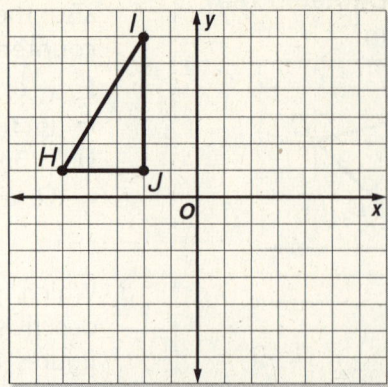

△ HIJ	After 90° Clockwise Turn	After 180° Counterclockwise Turn
H(−5, 1)		
I(−2, 6)		
J(−2, 1)		

EXTENDED RESPONSE

6 Josie picked a flower that looked like the one below. She rotated the flower 270° clockwise about her thumb.

Part A

In the space below, draw the flower in its new position.

Part B

By how many degrees should Josie rotate the flower if she wants it to be upside down from its original position? Should she turn it clockwise or counterclockwise? Write and explain your answer on the lines below.

LESSON 6.8 Drawing an Image Under Reflection

New York Performance Indicators

8.G.9   Draw the image of a figure under a reflection over a given line

8.CM.4 Share organized mathematical ideas through the manipulation of objects, numerical tables, drawings, pictures, charts, graphs, tables, diagrams, models and symbols in written and verbal form

8.R.1   Use physical objects, drawings, charts, tables, graphs, symbols, equations, or objects created using technology as representations

8.R.11 Use mathematics to show and understand mathematical phenomena (e.g., use tables, graphs, and equations to show a pattern underlying a function)

VOCABULARY

Reflection is flipping a figure over a line in the plane

REVIEW

Understanding Reflections

A figure can be reflected or flipped over any line. The diagram below shows △*ABC* reflected over the *y*-axis.

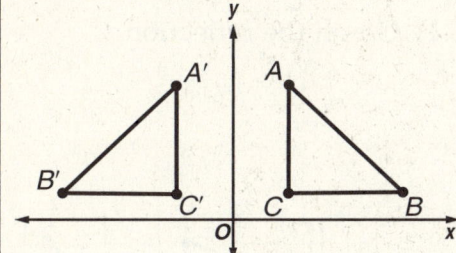

What You Should Know

A figure under reflection is the mirror image of the original figure. The line that the figure is flipped over is called the **line of symmetry**.

Line of Symmetry

Drawing Reflections Over the *x*- and *y*-axis

You can easily draw the reflection of a figure by applying the following rules to each vertex or given point.

<u>Reflection</u>	<u>Rule</u>
over the *x*-axis	Multiply the *y*-coordinate by -1: $(x, y) \longrightarrow (x, -y)$
over the *y*-axis	Multiply the *x*-coordinate by -1: $(x, y) \longrightarrow (-x, y)$

EXAMPLE 1

Graph the reflection of quadrilateral *DEFG* over the *y*-axis.

You can use the rule for reflections over the y-axis.

List the coordinates of each vertex in the figure and apply the rule for a reflection over the *y*-axis.

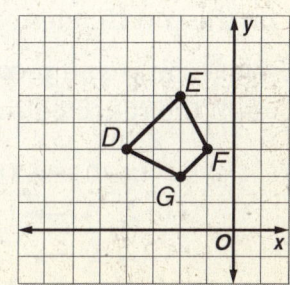

Multiply the *x*-coordinate by −1, and use the same *y*-coordinate.

$$(x, y) \longrightarrow (-x, y)$$
$$D(-4, 3) \longrightarrow D'(4, 3)$$
$$E(-2, 5) \longrightarrow E'(2, 5)$$
$$F(-1, 3) \longrightarrow F'(1, 3)$$
$$G(-2, 2) \longrightarrow G'(2, 2)$$

Use the new ordered pairs to graph the reflected image.

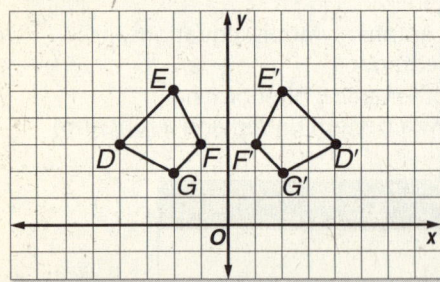

▶ **Understanding the Solution** To check your answer, make sure that corresponding points between the original figure and the image are the same distance from the *y*-axis.

TRY IT!

The vertices of △*QRS* are *Q*(1, 6), *R*(6, 4), and *S*(4, 1). Graph the reflection of △*QRS* over the *x*-axis.

Drawing Reflections Over the Line *y* = *x*

Look at the line *y* = *x* and the rule for reflecting a figure over this line. Apply the rule to each vertex or given point of the original figure.

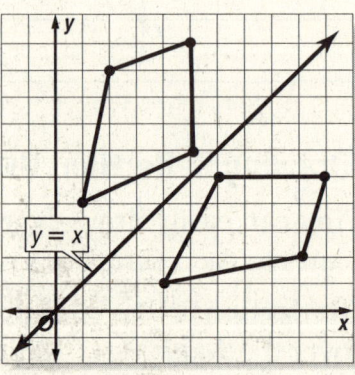

Reflection
over the line *y* = *x*

Rule
Switch the *x*- and *y*-coordinates:
$$(x, y) \longrightarrow (y, x)$$

EXAMPLE 2

Graph the reflection of △*FGH* over the line *y* = *x*.

You can use the rule for reflections over this line.

List the coordinates of each vertex in the figure and apply the rule for a reflection.

Switch the *x*- and *y*-coordinates.
Graph the reflected image.

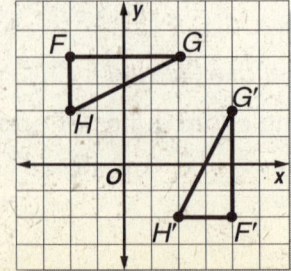

$$(x, y) \longrightarrow (y, x)$$
$$F(-2, 4) \longrightarrow F'(4, -2)$$
$$G(2, 4) \longrightarrow G'(4, 2)$$
$$H(-2, 2) \longrightarrow H'(2, -2)$$

▶ **Understanding the Solution** You can check your graph by folding the paper on the line $y = x$. The original figure and the image should line up. This works for any line of symmetry.

TRY IT!

Graph the reflection of figure *MNOP* over the line $y = x$.

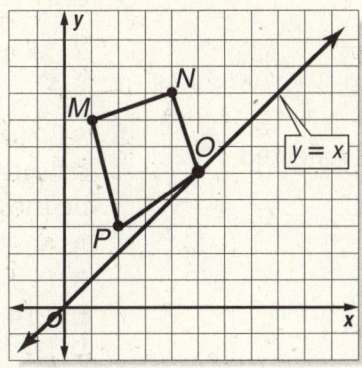

Exercises

SHORT RESPONSE

1 Reflect △*HIJ* over the *x*-axis. Then reflect △*H'I'J'* over the *y*-axis. Write the coordinates of the vertices for each image on the lines below.

△ *HIJ*	After Reflection Over *x*-axis	△ *H'I'J'* Reflected Over *y*-axis
H(1, −4)		
I(2, −1)		
J(4, −3)		

2 Graph both images from problem 1. What other transformation of △*HIJ* could result in the second image?

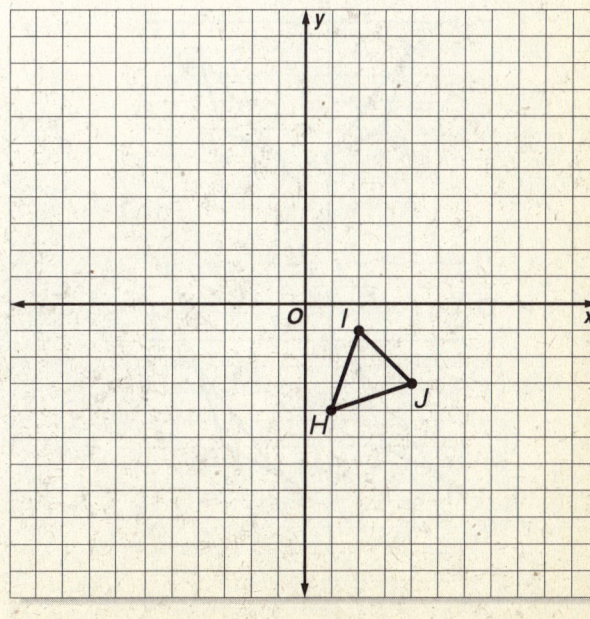

Use the graph below to answer questions 3 and 4.

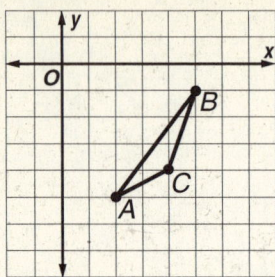

3 Which graph shows △*ABC* reflected over the *x*-axis?

A

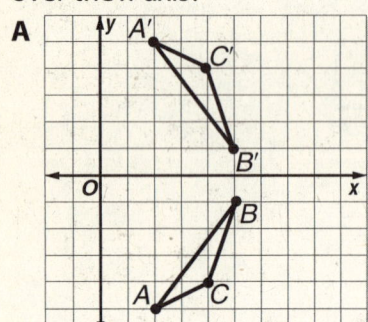

B

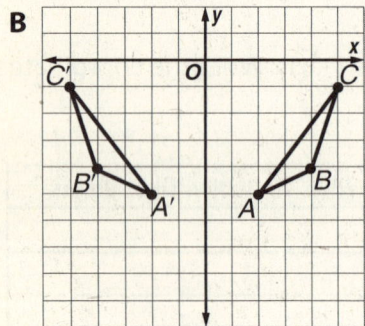

C

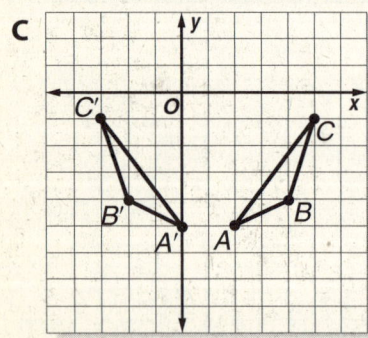

D

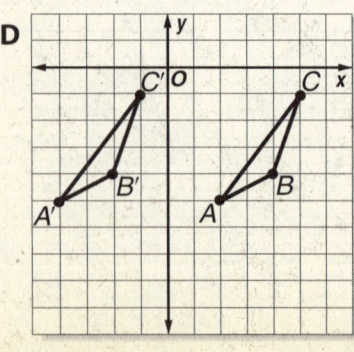

4 What would be the coordinates of *A'* if the triangle was reflected over the line *y = x*?

F (2, −5)

G (−5, 2)

H (−2, 5)

I (2, 5)

5 A picture with one point (−3, 4) was reflected over the *y*-axis. What should be the new coordinates of the point?

A (4, −3)

B (−3, −4)

C (3, 4)

D (3, −4)

6 What two reflections could result in the image *C'V'B'N'* shown below?

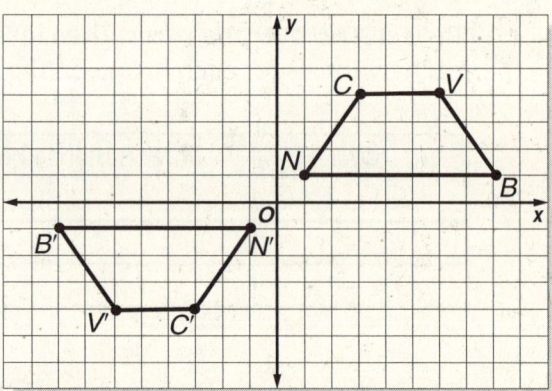

F reflection over the *x*-axis

G reflection over the *y*-axis

H reflection over the line *y = x*

J reflection over the *x*-axis and then over the *y*-axis

7 A point (3, 2) is reflected over the line *y = x*. That new point is then reflected over the *x*-axis. What are the coordinates of the newest point?

A (2, 3)

B (−2, 3)

C (2, −3)

D (−2, −3)

8 Look at the transformation of each figure below.

Part A

What is the line of reflection for each transformation below?

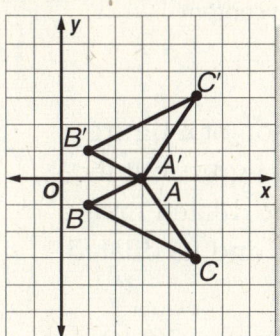   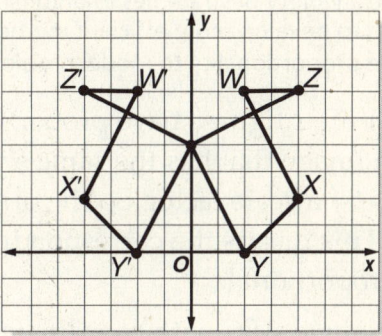

Answer _____   **Answer** _____

Part B

How does reflecting or rotating a figure change the interior angles of the figure?

_____

_____

_____

LESSON 6.9 · *Problem-Solving Strategy: Proportional Reasoning*

New York Performance Indicators

8.PS.1   Use a variety of strategies to understand new mathematical content and to develop more efficient methods

8.PS.10 Use proportionality to model problems

Recall that similar figures have proportional sides. A dilation of a figure results in an image that has the same shape as the original figure, but its size changes by a scale factor. So the original figure and the dilated image are similar. This means the corresponding sides of a figure and its dilated image are proportional.

Problem: Triangle ABC with coordinates at $A(-2, 2)$, $B(4, 2)$, and $C(2, 4)$ is dilated by a scale factor of $\frac{1}{2}$. Find the ratio of the corresponding sides of the original triangle and its image.

SOLUTION

What do you know?

The scale factor is $\frac{1}{2}$.

What do you need to find?

the ratio of corresponding sides

Strategy

Multiply the coordinates of the figure's vertices by the scale factor to get the vertices of the image. $A'(-1, 1)$, $B'(2, 1)$, and $C'(1, 2)$

To find the ratio of corresponding side lengths of the triangle and its image, you can use the scale factor. The sides of the dilated image are half the length of the sides of the original triangle.

$$A'B' = \tfrac{1}{2}AB \qquad B'C' = \tfrac{1}{2}BC \qquad C'A' = \tfrac{1}{2}CA$$

You can also use the distance formula $\sqrt{(x_2 - x_1)^2 + (y_2 - y_1)^2}$ to find the ratio of corresponding side lengths.

$$AB = \sqrt{(4 - (-2))^2 + (2 - 2)^2} \qquad A'B' = \sqrt{(2 - (-1))^2 + (1 - 1)^2}$$
$$= \sqrt{6^2} = 6 \qquad\qquad = \sqrt{3^2} = 3$$

So $\dfrac{A'B'}{AB} = \dfrac{3}{6} = \dfrac{1}{2}$.

You can express the ratio of corresponding sides as a proportion. Notice that the side lengths can be compared in either order.

$$\frac{A'B'}{AB} = \frac{B'C'}{BC} = \frac{C'A'}{CA} = \frac{1}{2} \text{ or } \frac{AB}{A'B'} = \frac{BC}{B'C'} = \frac{CA}{C'A'} = 2$$

▶ **Understanding the Solution** The second proportion says that the sides of the original triangle are twice as long as the sides of its dilated image.

EXTENDED RESPONSE

1 A square *SQRT* is dilated by a scale factor of 3. Write an equation that compares the lengths of $\overline{RT}$ and $\overline{R'T'}$. If the length of $\overline{Q'R'}$ is 6 cm, what is the perimeter of the original square?

Show your work.

Answer __________

2 A parallelogram *ABCD* is dilated by the scale factor 6. The image $A'B'C'D'$ is then dilated by a scale factor of $\frac{1}{3}$. Write an equation that relates the length of $\overline{AB}$ of the original figure and the length of $\overline{A''B''}$ of the final image.

Show your work.

Answer ________________

LESSON 6.10 Drawing an Image Under Translation

New York Performance Indicators

8.G.10 Draw the image of a figure under a translation

8.PS.6 Represent problem situations verbally, numerically, algebraically and graphically

8.CM.4 Share organized mathematical ideas through the manipulation of objects, numerical tables, models, and symbols in written and verbal form

8.R.11 Use mathematics to show and understand mathematical phenomena (e.g., use tables, graphs, and equations to show a pattern underlying a function)

VOCABULARY

Translation is sliding a figure to a different location in the plane.

REVIEW

Understanding Translations

You can slide a figure up, down, and across any number of units. The diagram below shows one translation of △ABC.

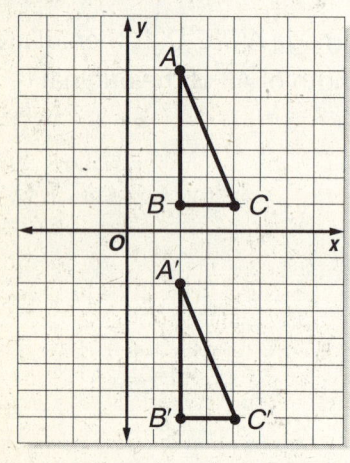

What You Should Know

In a translation, each point of a figure is moved the same distance and in the same direction.

A figure can be moved diagonally by sliding the figure both horizontally (right or left) and vertically (up or down). Look at the translation of point D.

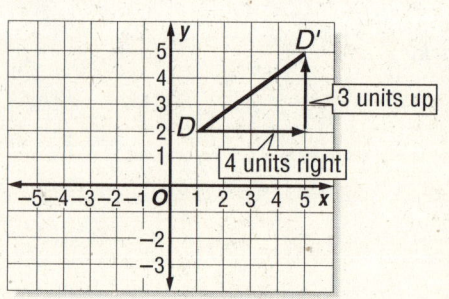

Drawing Translations

You can use an ordered pair (a, b) to describe a translation. The first number a is the number of units the figure is moved right or left. The second number b is the number of units the figure is moved up or down.

Direction	Rule
left or **down**	Write the units as a **negative** number.
right or **up**	Write the units as a **positive** number.

Using the rules above, a translation 2 units to the left and 3 units up can be described as (−2, 3). This ordered pair describes the change in the x- and y-values resulting from a translation. Adding this ordered pair to the coordinates for each vertex or point in the original figure results in the corresponding coordinates for the translated image.

EXAMPLE 1

△*JKL* has vertices *J*(−5, 1), *K*(−2, 2), and *L*(−1, 4). Graph the image of △*JKL* after sliding each vertex 5 units to the right and 3 units down. What are the coordinates of each of the vertices of △*J'K'L'*?

You can solve this problem using an ordered pair to describe the translation.

Write an ordered pair to describe the translation.

(5, −3)

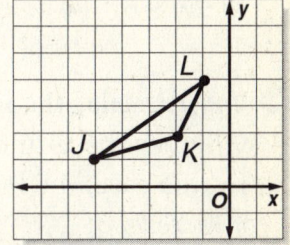

Add the first number to the *x*-coordinate and the second number to the *y*-coordinate of each vertex. Then use the new coordinates to draw △*J'K'L'*.

J(−5, 1) ⟶ (−5 + 5, 1 + [−3]) ⟶ *J'*(0, −2)
K(−2, 2) ⟶ (−2 + 5, 2 + [−3]) ⟶ *K'*(3, −1)
L(−1, 4) ⟶ (−1 + 5, 4 + [−3]) ⟶ *L'*(4, 1)

▶ Understanding the Solution You are adding 5 to each *x*-coordinate and adding −3 to each *y*-coordinate. The notation (*x*, *y*) ⟶ (*x* + 5, *y* − 3) helps you understand what you did to each of the points on the vertices to translate the image.

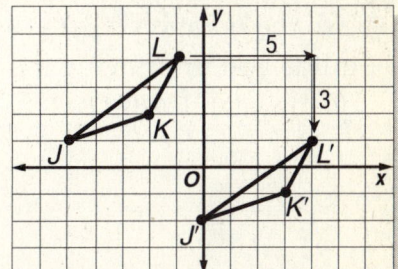

△*DEF* has vertices *D*(0, 2), *E*(3, 4), and *F*(5, 1). Translate △*DEF* 3 units to the left and 4 units up. What are the coordinates of points *D'*, *E'*, and *F'*?

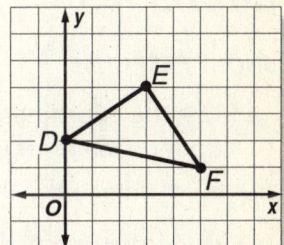

EXAMPLE 2

The vertices of a square *HIJK* are *H*(−6, −3), *I*(−3, −3), *J*(−3, −6), and *K*(−6, −6). Graph the original square *HIJK* and its image after a translation 4 units to the right and 2 units up.

*You can solve this problem using the notation (*x*, *y*) ⟶ (*x* + 4, *y* + 2).*

Use the given coordinates to graph *HIJK*. Then use the notation describing the translation to find the coordinates of each new vertex, and graph *H'I'J'K'*.

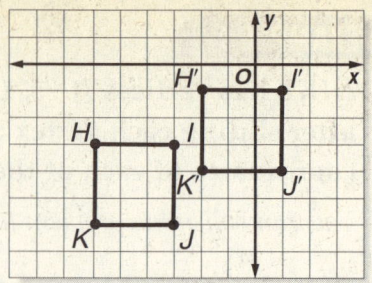

$H(-6, -3) \longrightarrow (-6 + 4, -3 + 2) \longrightarrow H'(-2, -1)$
$I(-3, -3) \longrightarrow (-3 + 4, -3 + 2) \longrightarrow I'(1, -1)$
$J(-3, -6) \longrightarrow (-3 + 4, -6 + 2) \longrightarrow J'(1, -4)$
$K(-6, -6) \longrightarrow (-6 + 4, -6 + 2) \longrightarrow K'(-2, -4)$

▶ **Understanding the Solution** The notation $(x, y) \longrightarrow (x + 4, y + 2)$ lets you know whether the *x*- and *y*-coordinates will increase or decrease and by how much. In this case, the image was translated in a positive direction, since both *x* and *y* increased.

TRY IT!

Use the notation $(x, y) \longrightarrow (x - 3, y - 1)$ to guide you in drawing the translation of quadrilateral *ABCD*.

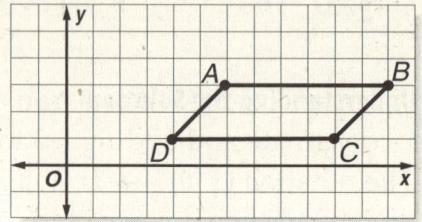

Exercises

SHORT RESPONSE

1 Use the coordinate grid to the right to graph △*PQR* under each of the two following translations.

a. 4 units to the left

b. 2 units to the right and 2 units up

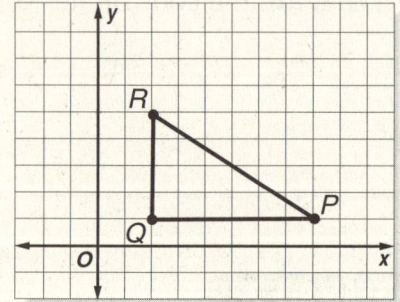

Use an ordered pair and arrow notation to describe each translation.

a: (_____ , _____)

notation: _____

b: (_____ , _____)

notation: _____

2 If point $D(-4, 5)$ is translated 3 units to the right and 4 units down, which coordinates represent D'?

A $(-7, 7)$

B $(-1, -3)$

C $(-1, 1)$

D $(1, 9)$

3 Which notation describes the translation of the figure shown below?

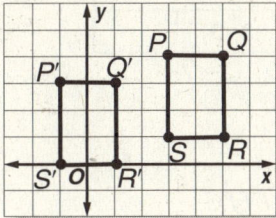

F $(x, y) \longrightarrow (x - 4, y - 1)$

G $(x, y) \longrightarrow (x + 3, y - 1)$

H $(x, y) \longrightarrow (x + 4, y + 1)$

J $(x, y) \longrightarrow (x - 2, y + 1)$

4 Which of the following describes the translation of $\triangle DEF$ below?

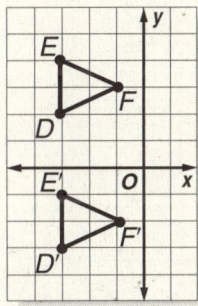

A slid 2 units to the right

B slid 5 units down

C slid 3 units to the left and 2 up

D slid 3 units to the right and 2 down

5 Which best describes the translation for this notation: $(x, y) \longrightarrow (x + 3, y - 1)$?

F translation 3 units to the left and 1 unit down

G translation 3 units to the right and 1 unit down

H translation 3 units to the right and 1 unit up

J translation 1 unit to the right and 3 up

6 Which graph shows the quadrilateral *QRST* after a translation of 1 unit to the left and 2 units up?

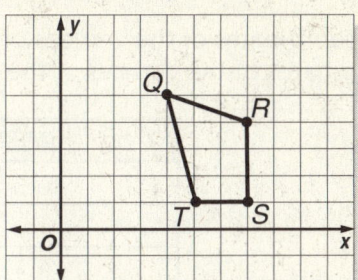

A

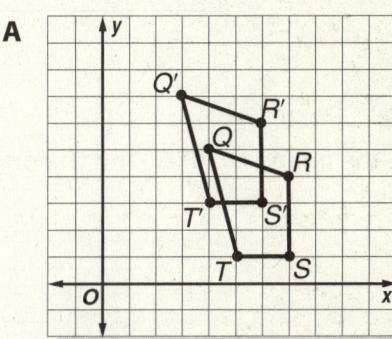

B

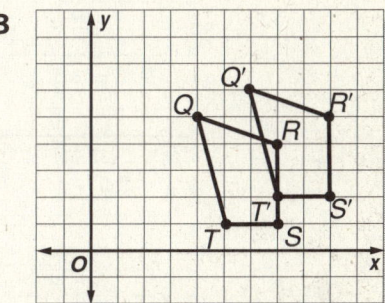

C

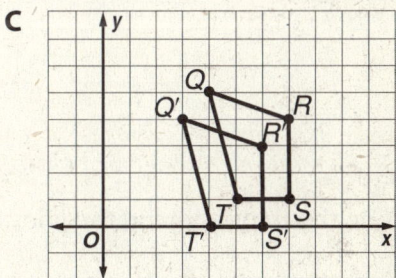

D

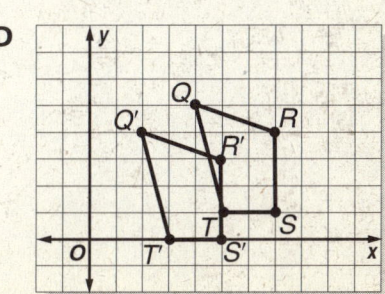

7 The transformation of quadrilateral *DEFG* is described by $(x, y) \longrightarrow (x+2, y-3)$.

Part A

The coordinates of the vertices of the image *D′E′F′G′* are given in the table. Find the coordinates of the vertices of the original quadrilateral *DEFG*.

DEFG	D′E′F′G′
	D′(−3, 3)
	E′(1, 3)
	F′(1, 0)
	G′(−3, 0)

Part B

Draw the image *D′E′F′G′* and the original quadrilateral, *DEFG*.

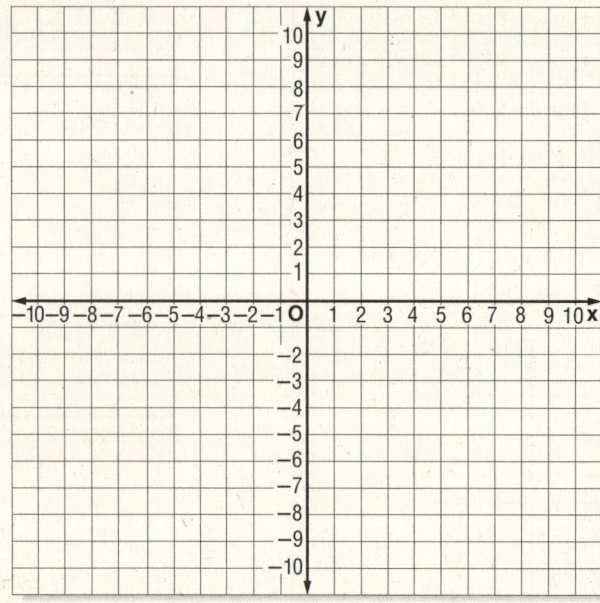

Part C

Describe the translation on the lines below.

Drawing an Image Under Dilation

New York Performance Indicators

8.G.11 Draw the image of a figure under a dilation

8.PS.6 Represent problem situations verbally, numerically, algebraically and graphically

8.CM.4 Share organized mathematical ideas through the manipulation of objects, numerical tables, models, and symbols in written and verbal form

8.R.11 Use mathematics to show and understand mathematical phenomena (e.g., use tables, graphs, and equations to show a pattern underlying a function)

VOCABULARY

A **dilation** is a transformation in which the size of a figure is changed, but not its shape. The dilated image is similar but not congruent to the original.

The **scale factor** determines how much larger or smaller a dilated image is than the original figure.

REVIEW

Understanding Dilated Images

You can make a figure larger or smaller without changing its shape. The graph below shows a dilation of $\triangle DEF$ by a scale factor of 2. Notice that $\triangle D'E'F'$ is twice as big as $\triangle DEF$.

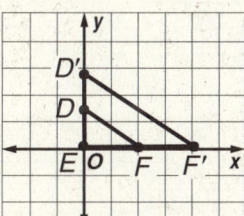

What You Should Know

Most dilations in the coordinate plane are centered about the origin, (0, 0).

For dilations of polygons, the scale factor r is the ratio of the length of one side of the dilated image to the corresponding side of the original figure.

For $\triangle DEF$ and $\triangle D'E'F'$, $r = \dfrac{D'F'}{DF}$.

Drawing Dilations on a Graph

You can dilate a figure by multiplying the coordinates of its vertices or other points by a scale factor. For example, the dilation of $\triangle DEF$ shown above can be described as $(x, y) \longrightarrow (2x, 2y)$, where the scale factor is 2. The following rules can also help you draw dilations of figures.

Scale factor r	Rule
$r > 1$	Increase the size of the figure.
$0 < r < 1$	Reduce the size of the figure.

After multiplying the coordinates of each given point by the scale factor, use the new coordinates to draw the dilated image.

EXAMPLE 1

Hector drew △*ABC* for a drafting class. He needs to make the triangle 3 times larger. Draw a dilated image of △*ABC* using (0, 0) as the center of the dilation. What are the coordinates of the vertices of △*A'B'C'*?

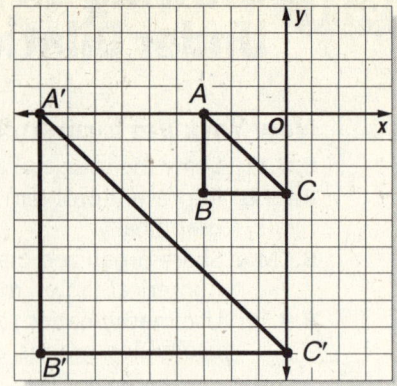

You can solve this problem using what you know about dilations.

Find the coordinates of the vertices of the original triangle:
A(−3, 0), *B*(−3, −3), and *C*(0, −3).

Multiply both numbers in each ordered pair by the scale factor, 3: *A'*(−9, 0), *B'*(−9, −9), and *C'*(0, −9). Use the new coordinates to graph the image.

▶ **Understanding the Solution** The dilated image has side lengths that are 3 times as long as the original. You can check your dilation by writing a ratio of corresponding sides: $\frac{B'C'}{BC} = \frac{9}{3} = 3$. Since 3 is the correct scale factor, your drawing is correct.

TRY IT!

△*JKL* has vertices *J*(0, 2), *K*(3, 4), and *L*(3, 1). If △*JKL* is dilated using a scale factor of 2 and (0, 0) as the center of the dilation, what will be the coordinates of points *J'*, *K'*, and *L'*?

EXAMPLE 2

Graph the dilated image of quadrilateral *ABCD* using a scale factor of 0.5 and (0, 0) as the center of dilation.

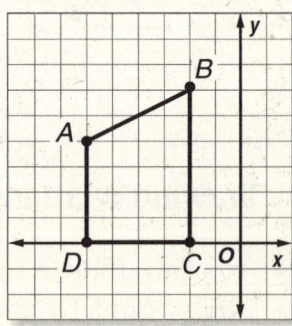

You can use the notation (x, y) ⟶ (0.5x, 0.5y) to guide you.

Find the coordinates of the vertices of the original figure: *A*(−6, 4), *B*(−2, 6), *C*(−2, 0), and *D*(−6, 0).

Multiply the numbers in each ordered pair by 0.5 or one half to find the coordinates of the translated image.

$$A(-6, 4) \longrightarrow (-6 \cdot 0.5, 4 \cdot 0.5) \longrightarrow A'(-3, 2)$$
$$B(-2, 6) \longrightarrow (-2 \cdot 0.5, 6 \cdot 0.5) \longrightarrow B'(-1, 3)$$
$$C(-2, 0) \longrightarrow (-2 \cdot 0.5, 0 \cdot 0.5) \longrightarrow C'(-1, 0)$$
$$D(-6, 0) \longrightarrow (-6 \cdot 0.5, 0 \cdot 0.5) \longrightarrow D'(-3, 0)$$

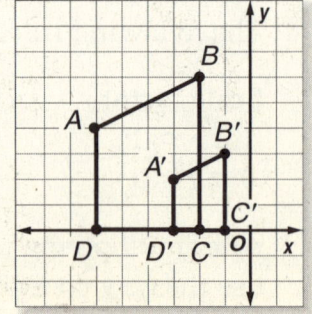

Use the new coordinates to graph the image.

▶ **Understanding the Solution** The notation $(x, y) \longrightarrow (0.5x, 0.5y)$ lets you know how much larger or smaller the dilated image will be. In this case, the side lengths of the image are half (0.5 times) the side lengths of the original.

TRY IT!

Use the notation $(x, y) \longrightarrow (0.25x, 0.25y)$ to guide you in drawing the dilation of $\triangle CDE$. Label the vertices of the image using prime notation.

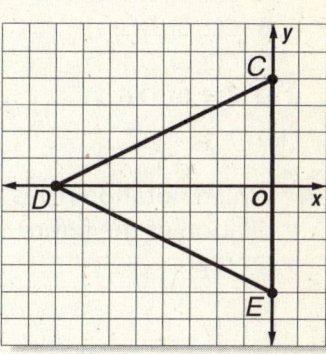

Exercises

SHORT RESPONSE

1 Graph the dilation of $\triangle QRS$ using a scale factor of 1.5 and the origin as the center of dilation. Use arrow notation to describe the dilation, and list the coordinates of the image.

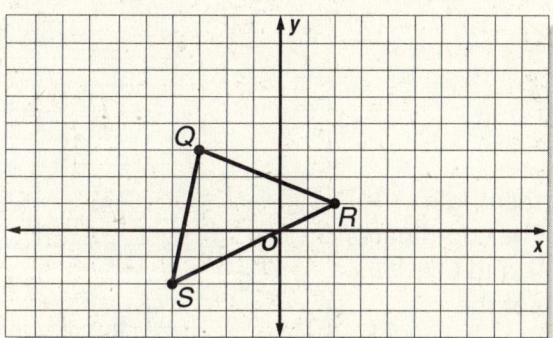

$(x, y) \longrightarrow$ _____

△ QRS	△ Q′R′S′
Q (−3, 3)	Q′ _____
R (2, 1)	R′ _____
S (−4, −2)	S′ _____

2 In question 1, how many times longer or shorter is side $Q'R'$ compared to side QR?

3 Line segment *CD* was dilated around the origin by a factor of 2. The endpoints of the image are *C*′(4, 0) and *D*′(6, 2). What are the coordinates of the endpoints of the original line segment?

A *C*(2, 0), *D*(3, 0)

B *C*(2, 0), *D*(3, 1)

C *C*(2, 0), *D*(1, 1)

D *C*(4, 0), *D*(6, 2)

4 Which notation gives the rule for the dilation of the figure *MNOP* shown?

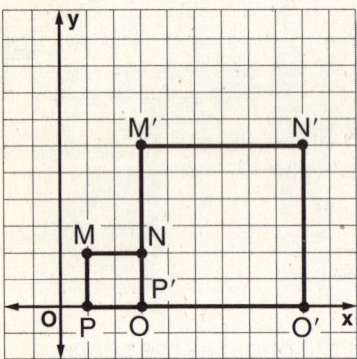

F $(x, y) \longrightarrow (3x, 3y)$

G $(x, y) \longrightarrow (0.5x, 0.5y)$

H $(x, y) \longrightarrow (2x, 2y)$

J $(x, y) \longrightarrow (0.25x, 0.25y)$

5 △*ABC* was dilated about the origin using a scale factor of 3. The coordinates of △*ABC* are *A*(0, 2), *B*(3, 4), and *C*(3, 1). What are the coordinates of the dilated image?

A *A*′(3, 5); *B*′(6, 7); *C*′(6, 4)

B *A*′(3, 6); *B*′(9, 12); *C*′(9, 3)

C *A*′(0, 6); *B*′(9, 12); *C*′(9, 0)

D *A*′(0, 6); *B*′(9, 12); *C*′(9, 3)

6 The dilation of a figure can be described as $(x, y) \longrightarrow (0.75x, 0.75y)$. What is the size of the dilated image with respect to the original figure?

F The side lengths are 75 times longer.

G The side lengths are 0.75 times as long.

H The side lengths are increased by 0.75 units.

J The side lengths are decreased by three-fourths.

7 Which graph shows △*DEF* after a dilation by a scale factor of 2, using the origin as the center of dilation?

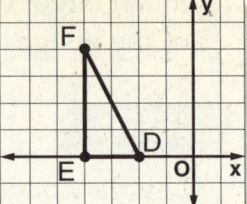

A

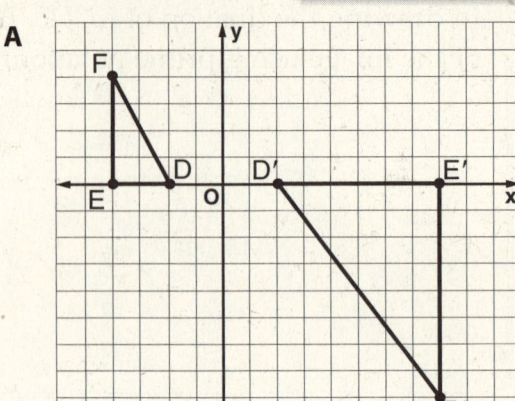

B

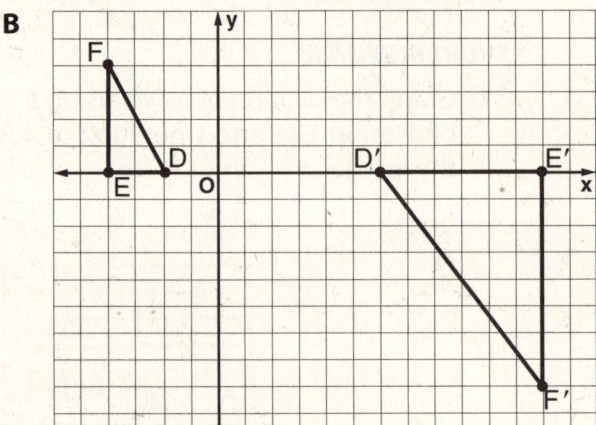

C

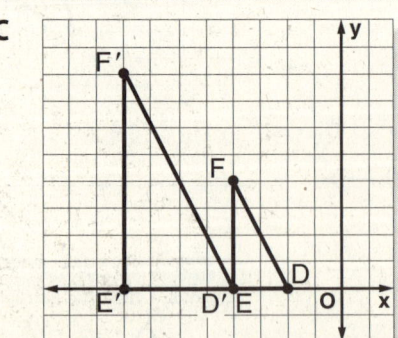

D

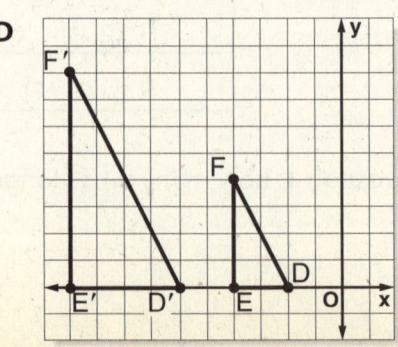

8 Myles was asked to draw a triangle *XYZ* and then dilate it using the origin as the center of dilation and a scale factor of 0.75. He thinks that adding 0.75 units to the coordinates of the vertices of △*XYZ* will help him draw the dilated image.

Part A

Explain why Myles is incorrect, and what he needs to do to graph the dilation.

 Use your ruler to help you solve this problem.

Part B

Use the coordinate plane below to draw a triangle *XYZ*. Then draw its dilated image using the origin as the center of dilation and a scale factor of 0.75. Make sure to label the vertices of each triangle.

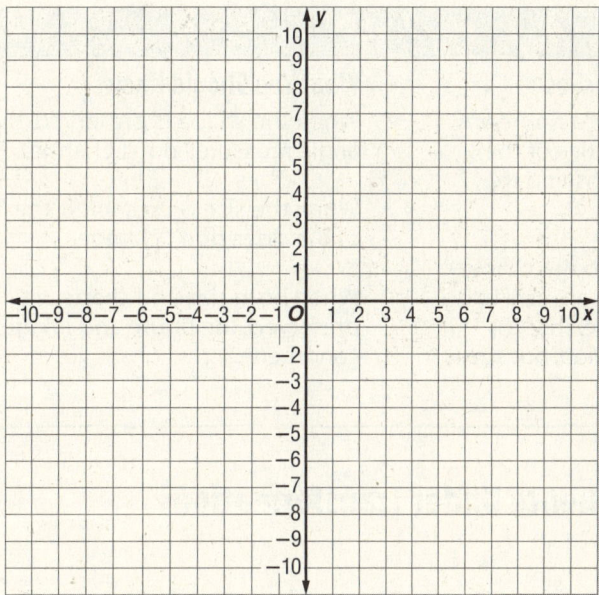

LESSON 6.12 — Identifying Properties Under Transformation

New York Performance Indicators

8.G.12 Identify the properties preserved and not preserved under a reflection, rotation, translation and dilation

8.RP.1 Recognize that mathematical ideas can be supported by a variety of strategies

8.CM.10 Use appropriate language, representations, and terminology, when describing objects, relationships, mathematical solutions, and rationale

8.R.11 Use mathematics to show and understand mathematical phenomena (e.g., use tables, graphs, and equations to show a pattern underlying a function)

VOCABULARY

The **properties** of a figure are its shape and size that are related to its side lengths and angle measures.

Congruent figures have the same size and shape.

Similar figures have the same shape but can be different sizes. The lengths of corresponding sides are proportional.

REVIEW

Understanding Properties

The shape and size of a figure are related to the properties of the figure: the side lengths and the angle measures.

The properties that do not change under transformation are said to be *preserved*. For example, shape is preserved under dilation, but size is not preserved.

What You Should Know

When the shape is preserved, the angle measures don't change.

When the size is preserved, the lengths of the sides don't change.

When both size and shape are preserved, the figure and image are congruent.

Identifying Congruence Under Transformations

EXAMPLE 1

Compare △J′K′L′ to △JKL. Identify the transformation(s) that resulted in the image. Are the figure and its image congruent?

You can solve this problem using what you know about transformations.

The shape is preserved but the size is not preserved, so the figure has been dilated. This means that the corresponding side lengths of the two figures are not congruent.

You can use the distance formula to show that the side lengths are not equal.

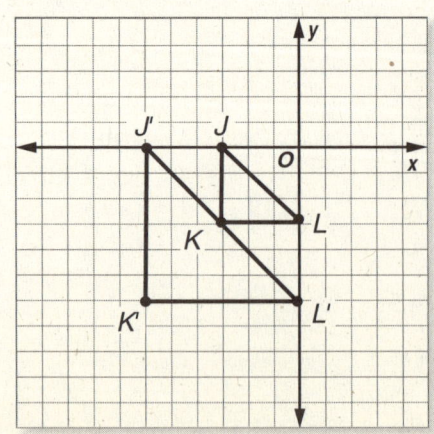

Side of triangle	$\overline{JL}$	$\overline{J'L'}$
Coordinates of endpoints	$J(-3, 0)$ and $L(0, -3)$	$J'(-6, 0)$ and $L'(0, -6)$
Length of side	$d = \sqrt{(-3-0)^2 + (0-(-3))^2}$	$d = \sqrt{(-6-0)^2 + (0-(-6))^2}$
	$= \sqrt{9+9}$	$= \sqrt{36+36}$
	$= \sqrt{18}$	$= \sqrt{72}$

$\sqrt{18} \neq \sqrt{72}$, so the triangles are not congruent.

Next check for orientation. Write the vertices of the figure and its image in clockwise-order: JLK and $J'L'K'$. Since the order of the vertices is the same, the orientation of the triangle and its image doesn't change under dilation.

▶ **Understanding the Solution** Dilations produce figures that are similar but not congruent. You can use the side lengths you found to determine the scale factor of the dilation: $\dfrac{J'L'}{JL} = \dfrac{\sqrt{72}}{\sqrt{18}}$ $\dfrac{\sqrt{72}}{\sqrt{18}} = \sqrt{4} = 2$. $\triangle J'K'L'$ is twice the size of $\triangle JKL$.

TRY IT!

Compare $\triangle A'B'C'$ to $\triangle ABC$. Identify which properties of $\triangle ABC$ were preserved and which transformation(s) took place.

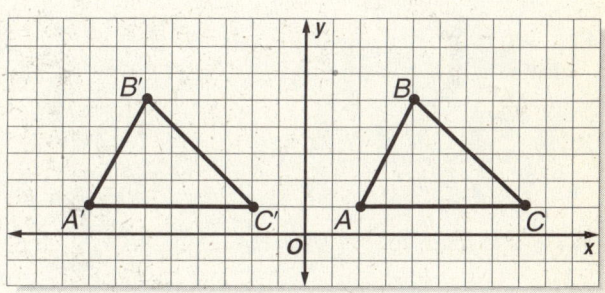

EXAMPLE 2

The vertices of a figure $ABCD$ are $A(0, 4)$, $B(4, 6)$, $C(4, 0)$, and $D(0, 0)$. If the figure is reflected over the y-axis, will the resulting image be congruent?

You can solve this problem by using the definition of reflection.

A reflection results in a mirror image of a figure across the line of reflection. The location and orientation of the figure are changed, but the size and shape of the figure and its image are the same, or congruent.

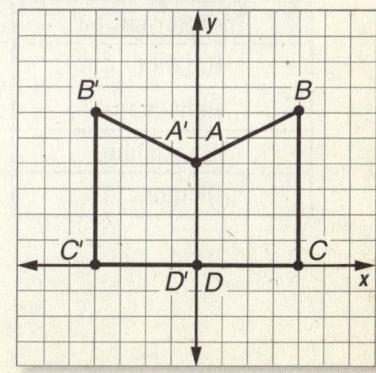

▶ **Understanding the Solution** To check your answer, you can graph the original figure and its reflection. Writing the vertices of the original figure and the image in clockwise-order results in $ABCD$ and $A'D'C'B'$. The orientation of the figure has changed under reflection.

If △*ABC* is rotated 180° about the origin, will the resulting image △*A′B′C′* be congruent? What properties of △*ABC* are preserved under rotation?

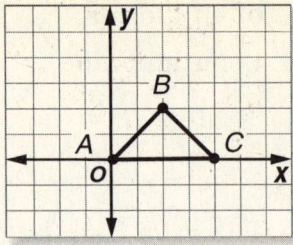

Exercises

SHORT RESPONSE

1 Transform *DEFGH* to produce a congruent figure. Graph and label your image on the coordinate grid.

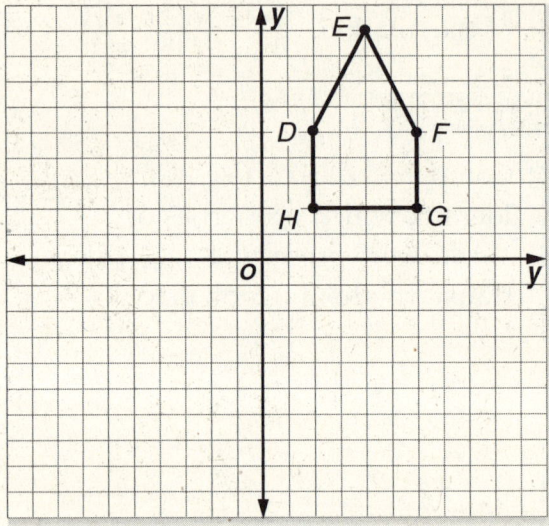

Explain how you know your image is congruent to *DEFGH* on the lines below.

_____

_____

2 Complete the table below by marking the properties that change under each transformation with an "X."

Transformation	Size	Shape	Location in the Plane	Orientation
Rotation				
Reflection				
Translation				
Dilation				

3 A rotation transforms an image by turning it around a point. What property is changed in the rotated image?

   A length of sides

   B measure of angles

   C location of figure

   D shape of figure

4 This notation shows what was done to transform a figure on a coordinate grid: $(x, y) \longrightarrow (0.5x, 0.5y)$. What is true about the resulting image?

   F The image is congruent to the original.

   G The image is translated 0.5 units to the left and down.

   H The orientation of the image has changed.

   J The image is similar but not congruent to the original.

5 Which of the following transformations does not result in a congruent figure?

   A reflection

   B dilation

   C rotation

   D translation

6 A figure is dilated using the origin as the center of dilation and a scale factor of 4. Which property of the original figure has been changed?

   F size

   G location

   H shape

   J orientation

7 Which is true about a translation?

   A The angles inside the figure change.

   B The side lengths of the figure change.

   C The image and the original figure are congruent.

   D The orientation of the figure changes.

8 Which properties of the original figure have changed in the transformation shown below?

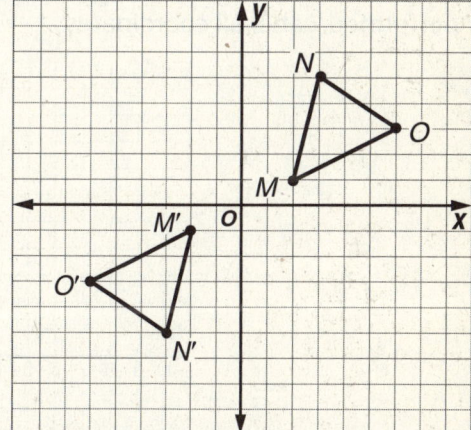

   F size and shape

   G size and location

   H location and orientation

   J orientation and shape

EXTENDED RESPONSE

9 A triangle *ABC* with vertices at *A*(2, –2), *B*(2, 3), and *C*(–4, –2) is reflected over the *x*-axis, rotated 90° clockwise about the origin, and then translated 3 units down and 2 units left.

Part A

Describe the operations you must perform on the coordinates of the vertices for each transformation listed above. List the coordinates of the new vertices after each transformation.

Reflection over x-axis:

_____

Rotation 90° clockwise about the origin:

_____

Translation:

_____

Part B

Graph both the original triangle and the final image after all transformations have been performed. Label all coordinates.

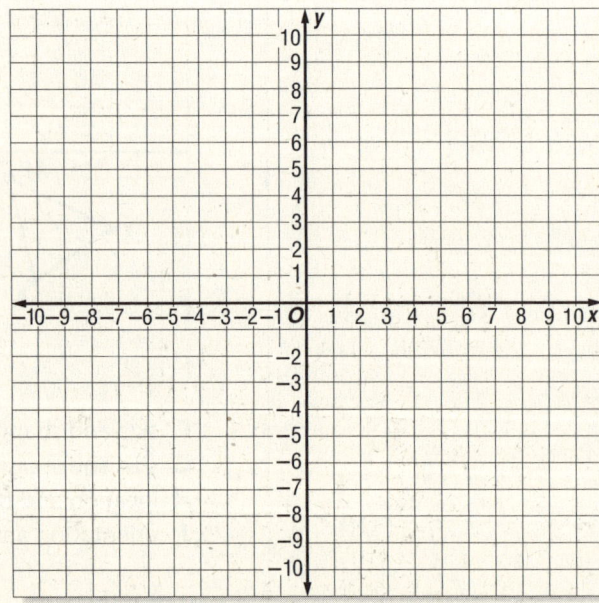

Part C

Is the new triangle congruent to the original triangle? Explain why or why not on the lines below.

_____

_____

_____

Problem-Solving Strategy: Applying Transformations

 New York Performance Indicators

8.PS.2 Construct appropriate extensions to problem situations

Understand the Strategy

If a car moves at a constant or uniform speed v, then distance s is a linear function of time t: $s(t) = vt$. You can use a transformation to write the distance function. Suppose the car traveled a certain distance d before reaching the constant speed. Then the total distance is given by the function $s(t) = vt + d$.

Problem: Jerry stopped at the supermarket on his way home from work. The supermarket is 5 miles from his office. From the supermarket, he drove at a constant speed of 45 miles per hour and arrived home 30 minutes later. How many miles did Jerry travel altogether?

SOLUTION

What do you know?

office to supermarket: 5 miles

supermarket to home:
30 min. at $v = 45$

What do you need to find?

the distance from Jerry's office to the supermarket to his home

Find the relationship.

Distance after leaving the supermarket: $\longrightarrow s(t) = 45t$

Distance between work and the supermarket: $\longrightarrow$ 5 miles

Total distance between work and home: $\longrightarrow s(t) = 45t + 5$

Change the unit of time from minutes to hours. $\longrightarrow$ 30 minutes = 0.5 hour

Find Jerry's total distance from work by substituting the time into the function.
$$\longrightarrow s(t) = 45(0.5) + 5$$
$$= 27.5$$

▶ **Understanding the Solution** By translating the graph of $s(t) = 45t$ five units up, you can find the graph of $s(t) = 45t + 5$.

1 The speed *V* at which a ball travels after it is dropped is a linear function of time *t*.

$$V(t) = -9.8t$$

Write a new function that describes the speed of the ball if it is thrown with an initial speed of 2 m/s. Describe the graph of the new function in terms of a transformation.

Show your work.

Answer _____

2 Carla has two bakeries. At both bakeries, the cost of making each muffin is $0.40. However, the fixed cost at the two locations is different. The graphs below show the cost functions at the two locations. Write the cost function for each graph. Then describe the relationship between the two graphs on the lines below.

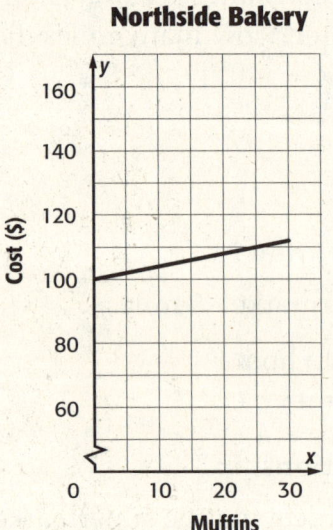

Eastside Bakery

Northside Bakery

Eastside Bakery _____   *Northside Bakery* _____

_____

_____

_____

MULTIPLE CHOICE

1 What type of transformation is shown below?

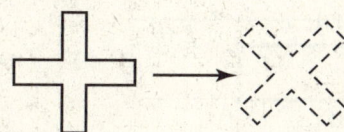

A reflection
B rotation
C translation
D dilation

2 Angles *A* and *B* are supplementary to each other. Angle *A* measures $(2x + 25)°$ and Angle *B* measures $(x + 5)°$. What is the value of *x*?

F 25
G 35
H 45
J 50

3 Figure *A* is reflected over the *x*-axis, translated 6 units in the positive *x* direction, and rotated ninety degrees clockwise to produce Figure *B*. What is the relationship between Figure *A* and Figure *B*?

A congruent but not similar
B similar but not congruent
C both similar and congruent
D neither similar nor congruent

4 What is the measure of angle 1 in the figure below?

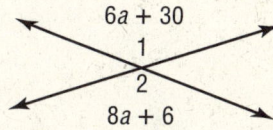

F 12°
G 72°
H 102°
J 110°

5 Which of the following statements is true when parallel lines are cut by a transversal?

A Same-side interior angles are congruent.
B Alternate exterior angles are supplementary.
C Corresponding angles are complementary.
D Alternate interior angles are congruent.

6 Using the figure, find *x*.

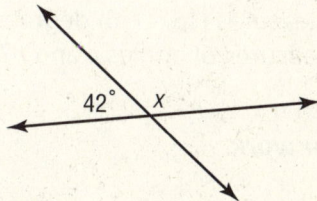

F 42°
G 48°
H 138°
J 142°

7 On the grid below, draw and label the reflection of rectangle *ABCD* over the line $y = x$.

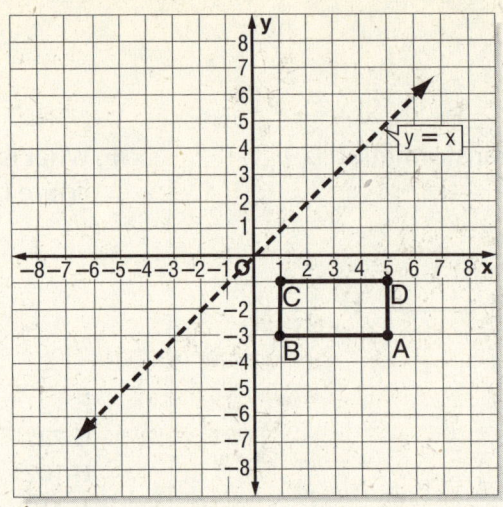

List the new coordinates for the vertices of the rectangle you just drew.

A′ (_____ , _____)   *C′* (_____ , _____)

B′ (_____ , _____)   *D′* (_____ , _____)

8 Parallel lines ℓ and m are cut by a transversal.

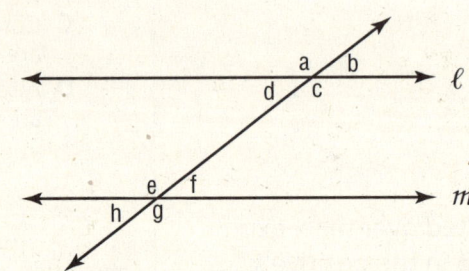

Angle *a* measures $(15x + 6)$ degrees, and $\angle f$ measures $(5x + 14)$ degrees. What are the degree measures of angles *c* and *h*?

Show your work.

m∠c _____   *m∠h* _____

LESSON 7.1 Converting Metric and Customary Measurements

8.M.1 Solve equations/ proportions to convert to equivalent measurements within metric and customary measurement systems

8.PS.10 Use proportionality to model problems

8.PS.13 Choose methods for obtaining required information

8.CN.6 Recognize and provide examples of the presence of mathematics in their daily lives

VOCABULARY

Customary measurement is the measurement system used in the United States.

Metric units are used in most other countries and by scientists all over the world. They include kilograms, meters, and liters.

REVIEW

Understanding Measurement Conversion

You can use ratios to convert between different measurements.

Examples:

20 feet is the same as 240 inches.

$20 \text{ feet} \times \dfrac{12 \text{ inches}}{1 \text{ foot}} = 240 \text{ inches}$

30 inches is the same as 2.5 feet.

$30 \text{ inches} \times \dfrac{1 \text{ foot}}{12 \text{ inches}} = 2.5 \text{ feet}$

What You Should Know

You can also set up a proportion for more complicated conversions.

When you multiply, the units in the numerator of one fraction should cancel out the units in the denominator of the other fraction.

$\dfrac{12 \text{ inches}}{1 \text{ foot}} \times \dfrac{3 \text{ feet}}{1 \text{ yard}} = \dfrac{36 \text{ inches}}{1 \text{ yard}}$

Customary Units of Length	Metric Units of Length
1 foot (ft) = 12 inches (in.)	1 centimeter (cm) = 10 millimeters (mm)
1 yard (yd) = 3 feet	1 meter (m) = 100 centimeters
1 mile (mi) = 5,280 feet	1 kilometer (km) = 1,000 meters
Customary Units of Weight	**Metric Units of Weight**
1 cup (c) = 8 fluid ounces (fl oz)	1 gram (g) = 1,000 milligrams (mg)
1 quart (qt) = 4 cups	1 kilogram (kg) = 1,000 grams
1 gallon (gal) = 4 quarts	
Customary Units of Capacity	**Metric Units of Capacity**
1 pound (lb) = 16 ounces (oz)	1 liter (L) = 1,000 milliliters (mL)
1 ton (T) = 2,000 pounds	

Converting Between Measurements

The table shows some common customary and metric conversions that you should know.

To convert from larger to smaller units (like from feet to inches), **multiply** by the conversion factor or number of smaller units per larger unit.

To convert from smaller to larger units (like from inches to feet), **divide** by the conversion factor, or multiply by the reciprocal.

EXAMPLE 1

Liz has a shelf that measures 62 inches long. Her grandmother has a 2-yard length of cloth she wants to give her to cover the shelf. Will the cloth be long enough?

You can solve this problem by converting from inches to yards.

You must convert from smaller to larger units. To find the number of inches in a yard, multiply the two conversion factors.

$$\frac{12 \text{ inches}}{1 \text{ foot}} \times \frac{3 \text{ feet}}{1 \text{ yard}} = \frac{36 \text{ inches}}{1 \text{ yard}}$$

Now you can set up a proportion. $\frac{62 \text{ inches}}{x \text{ yards}} = \frac{36 \text{ inches}}{1 \text{ yard}}$

Find the cross products and solve for x. $62 = 36x$ and $x = \frac{62}{36} \approx 1.7$.

The shelf is approximately 1.7 yards long, so 2 yards of cloth is enough to cover it.

▶ **Understanding the Solution** You could also have used the following proportion to convert 2 yards to inches: $\frac{x \text{ in.}}{2 \text{ yd}} = \frac{36 \text{ in.}}{1 \text{ yd}}$, so $x = 72$. The cloth is 72 inches long, which is long enough to cover a 62-inch shelf.

TRY IT!

Giselle has a pitcher that will hold 60 fluid ounces. Will this be enough to hold the 2 quarts of juice she needs for her party?

EXAMPLE 2

A penny has a diameter of about 2 centimeters. How many pennies could you line up end-to-end along a city block (about 1 kilometer)?

You can solve this problem by converting kilometers to centimeters.

One kilometer equals 1,000 meters, and 1 meter equals 100 centimeters.

$$\frac{1 \text{ kilometer}}{1,000 \text{ meter}} \times \frac{1 \text{ meter}}{100 \text{ centimeter}} = \frac{1 \text{ kilometer}}{100,000 \text{ centimeter}}$$

Now divide the number of centimeters in a kilometer by 2, the diameter of 1 penny: $100,000 \div 2 = 50,000$, so you could fit 50,000 pennies or $500.

▶ **Understanding the Solution** Metric units are easy to use because they are powers of 10.

$$\frac{1 \text{ km}}{10^3 \text{ m}} \times \frac{1 \text{ m}}{10^2 \text{ cm}} = \frac{1 \text{ km}}{10^5 \text{ cm}}$$

TRY IT!

One mosquito weighs about a milligram. If a medical research group collects about a kilogram of mosquitoes for a study on malaria, about how many mosquitoes did they collect?

Measurements for Area and Volume

Square units are always used in area calculation, while cubic units are used for volume.

EXAMPLE 3

Enrique's kitchen floor is 10 feet by 15 feet. He wants to cover the floor with either 1-foot by 1-foot tiles that cost $2 per square foot, or tiny 1-inch by 1-inch mosaic tiles that are sold for $0.02 a square inch. Which type of tile would be less expensive?

You can solve this problem by converting square feet to square inches.

Find the area of the kitchen floor: 10 feet $\times$ 15 feet = 150 square feet.

The cost of the large tiles would be 150 square feet $\times \dfrac{\$2}{1 \text{ square foot}} = \$300$.

There are 144 square inches in a square foot. To find the number of mosaic tiles that would be needed in the kitchen, use a proportion.

$$\frac{150 \text{ square feet}}{x \text{ square inches}} = \frac{1 \text{ square foot}}{144 \text{ square inches}}, \text{ and } 150 \times 144 = 1x \text{ or } x = 21,600$$

The cost of the mosaic tiles would be 21,600 square inches $\times \dfrac{\$0.02}{1 \text{ square inch}} = \$432$.

The mosaic tiles would cost more.

▶ **Understanding the Solution** To understand why there are 144 in^2 in a foot, imagine a 1-foot-by-1-foot square. Each side is 12 in. long, so the area of the square is 12 in. $\times$ 12 in. = 144 in^2.

TRY IT!

How many square feet are in a square yard? Use your answer to find the cost of 270 square feet of cloth at $4 a square yard.

Exercises

SHORT RESPONSE

1 A blue whale weighs approximately 100 tons, while a house cat weighs about 5 pounds. About how many cats would it take to equal the weight of one blue whale?
Show your work.

_____ cats

2 Convert 840 centimeters to meters.

A 8.4 meters

B 84 meters

C 8400 meters

D 84,000 meters

3 A road sign reads, "Roadwork ahead, 500 feet." How many miles will this be on the car odometer?

F 0.09

G 0.9

H 10.56

J 2,640,000

4 At birth, Jose Luis weighed 8.5 pounds. A bag of apples weighs 80 ounces. Which weighs more?

A The apples weigh more.

B Jose Luis weighs more.

C They both weigh the same.

D The answer cannot be determined from this information.

5 Convert 2.3 miles to inches.

F 27.6 in.

G 1012 in.

H 27,547 in.

J 145,728 in.

6 A dime weighs about 2.3 grams. If Alex has a collection of dimes that weighs nearly 1 kilogram, about how much money does he have?

A $434

B $43.40

C $230

D $2,300

7 A playground measures 18 yards by 22 yards. Find the area in square feet.

F 44 ft^2

G 396 ft^2

H 1,188 ft^2

J 3,564 ft^2

8 How many square feet are in a square mile?

A 9 ft^2

B 144 ft^2

C 5,280 ft^2

D 27,878,400 ft^2

9 How many **cubic** meters are in a **cubic** kilometer?

F 10 m^3

G 1,000 m^3

H 1,000,000 m^3

J 1,000,000,000 m^3

EXTENDED RESPONSE

10 A landscaper is ordering grass seed for a new city park. The total area of the park will be 250 acres. In order to grow a thick lawn, 3 pounds of grass seed must be planted for every 1,000 square feet.

Part A

One acre is equal to 43,560 square feet. How many pounds of grass seed will be needed to cover the entire area of the park?

Show your work.

Answer _____ pounds of seed

Part B

The grass seed supplier gives the city discounts for orders over 1 ton. The table below shows the prices for different amounts of Kentucky bluegrass seeds.

Tons of Seed	Price ($) per Ton
0.5–1	6,000
2–5	5,250
6–10	4,750
11–20	3,500
21 and over	3,000

How much will the grass seed cost for the new park?

Show your work.

Answer _____

MULTIPLE CHOICE

1 How many cubic inches are there in a cubic foot?

  A 12 cubic inches
  B 144 cubic inches
  C 1,728 cubic inches
  D 20,736 cubic inches

2 A truck is loaded up with boxes weighing a total of 272,000 ounces. How many tons is the truck carrying?

  F 8.5 tons
  G 17 tons
  H 136 tons
  J 17,000 tons

3 Which of the following belongs in the box?

  3.4 miles = ☐ inches

  A 1,795.2
  B 17,952
  C 21,542.4
  D 215,424

4 A vat of paint at a factory holds 84 gallons. The paint is being poured into tins that hold 384 fluid ounces each. How many tins can you fill with 1 vat of paint?

  F 3 tins
  G 8 tins
  H 28 tins
  J 48 tins

5 An airplane hangar measures 360 meters by 240 meters. Find its area in square kilometers.

  A 86,400 square kilometers
  B 864 square kilometers
  C 8.64 square kilometers
  D 0.0864 square kilometers

6 Gold costs about $13.50 per gram. The Bullion Depository at Fort Knox holds about 4.6 million kilograms of gold. About how much money is the gold in Fort Knox worth?

  F $62.1 million
  G $621 million
  H $6.21 billion
  J $62.1 billion

7 Nguyen has a deck of playing cards. He spreads out all 54 of the cards on top of a table, and he finds that they exactly cover the table from edge to edge. The table measures $2\frac{1}{2}$ feet long by $2\frac{1}{4}$ feet wide. What is the area of each playing card in square inches?

Show your work.

Answer _____ square inches

8 Jenny is filling up a shoebox with her marble collection. The shoebox is 12 inches long by 6 inches wide by 5 inches deep. Each marble takes up 1 cubic inch of space and weighs 8 grams. When the shoebox is totally full with marbles, how much will it weigh in kilograms?

Show your work.

Answer _____ kilograms

PART 1

1 Boris gets paid an 8% commission on every car he sells. Last week, he made a commission of $1,240. What was the price of the car he sold?

    A $15,500

    B $15,000

    C $12,500

    D $1,550

2 How many square centimeters are there in a square meter?

    F 100 square centimeters

    G 1,000 square centimeters

    H 10,000 square centimeters

    J 100,000 square centimeters

3 Simplify $(3x^3 + 2x^2 + 15) - (-2x^3 + 2x - 5)$.

    A $x^3 - 2x^2 - 2x + 10$

    B $x^3 + 2x^2 + 2x - 20$

    C $5x^3 + 2x^2 - 2x + 20$

    D $5x^3 - 10$

4 Adam, Chris, and June have to do the math homework they missed while they were sick. Adam has three more than twice the number of problems Chris has. However, he has fewer than 60 problems, the number that June must do. What inequality describes the number of problems that the three students must do?

F $x < 2x + 3 < 60$

G $x < 3 - 2x \leq 60$

H $3x + 3 < 60$

J $2x + 6 < 60$

5 Sheri has p pencils in her desk. Her friend Rosa has $4p - 5$ pencils. Which choice describes the number of pencils that Rosa has in her desk?

A five more than four times the number of pencils Sheri has

B five less than four times the number of pencils Sheri has

C four more than five times the number of pencils Sheri has

D four less than five times the number of pencils Sheri has

6 What is the product of $(3x + 7)$ and $(2x - 3)$?

F $6x^2 + 5x - 21$

G $6x^2 - 5x - 21$

H $x^2 + 5x + 21$

J $x^2 - 5x - 21$

7 Which is equivalent to the expression $(a^2 + a^4)^0$?

A 0

B 1

C a^6

D a^8

8 Which equation describes the graph below?

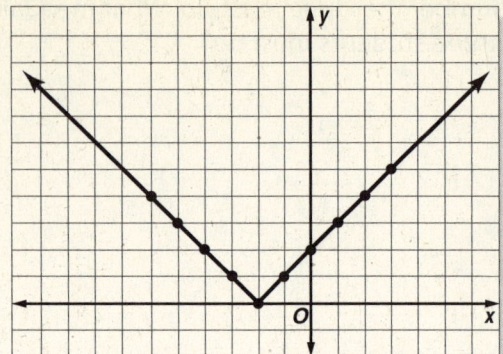

F $y = \frac{1}{2x} = 1$

G $y = (x + 2)^2$

H $y = |x + 2|$

J $y = -2$

9 The triangular park shown below is going to be surrounded by a fence. The dimensions of the park are given in feet. How many feet of fencing will be needed to surround the entire park?

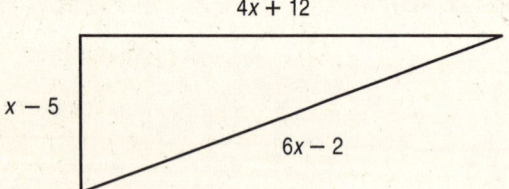

A $8x + 19$

B $11x + 5$

C $10x + 10$

D $11x - 19$

10 Simplify the expression $\frac{(x^4 \times x^3)^2}{x^4}$.

F x

G x^5

H x^{10}

J x^{20}

11 Find the solution to the inequality $-3x + 5 \leq -7$.

A $x \leq \frac{2}{3}$

B $x \geq -\frac{2}{3}$

C $x \leq 4$

D $x \geq 4$

12 The 8th-grade class at Pleasantville Middle School is going to run a booth called "Soak the Teacher" at the school fair. The materials for the booth cost $8, and they plan to charge $3 to every person who wants to play the game. Which graph shows the relationship between the number of people who play the game and the amount of profit the 8th graders make from their booth?

F
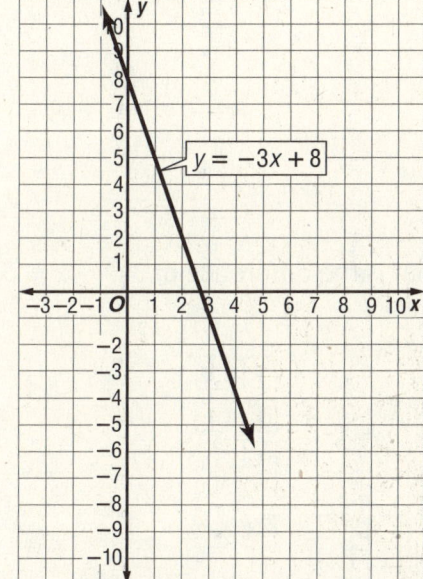
$y = -3x + 8$

G
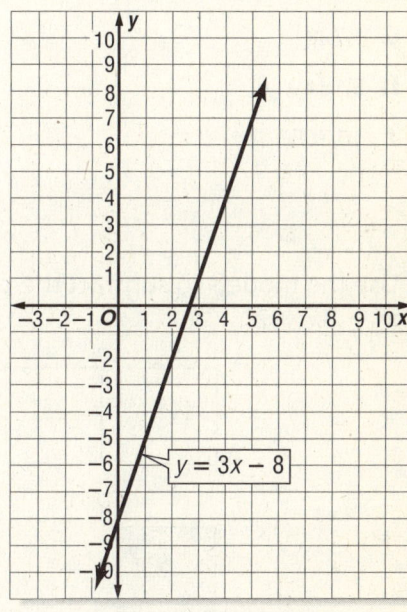
$y = 3x - 8$

H
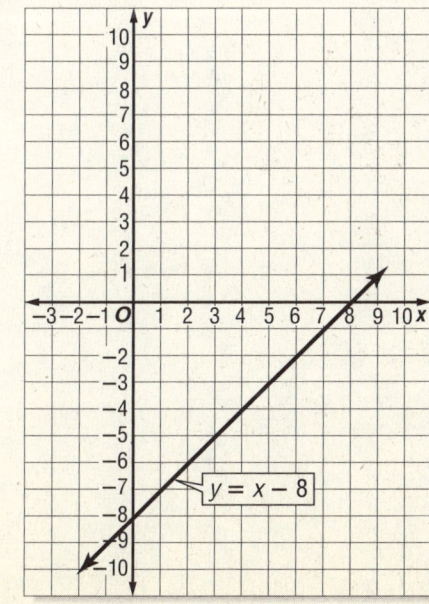
$y = x - 8$

J
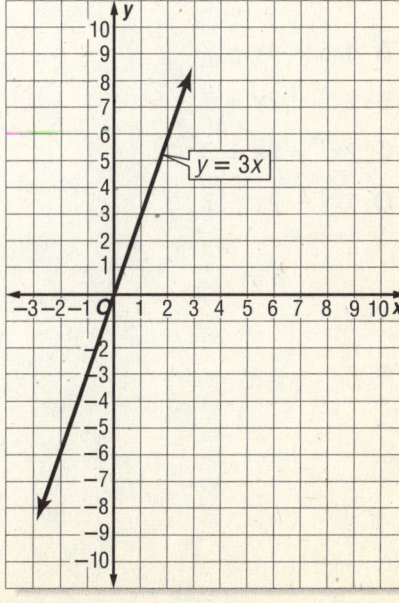
$y = 3x$

13 $\triangle ABC$ was dilated using the origin as the center of dilation and a factor of 2. The coordinates of the vertices of $\triangle A'B'C'$ are A' $(-4, 8)$, B' $(2, -2)$, and C' $(2, 8)$. What were the coordinates of the vertices of the original triangle?

A $A = (-8, 16)$, $B = (4, -4)$, $C = (4, 16)$

B $A = (2, -2)$, $B = (2, 8)$, $C = (-4, 8)$

C $A = (-2, 4)$, $B = (1, -1)$, $C = (1, 4)$

D $A = (2, 4)$, $B = (-1, -1)$, $C = (-1, 4)$

14 A crate full of sugar cubes has a volume of 3 cubic feet. The volume of each sugar cube is 0.5 cubic inch. How many sugar cubes are there in the crate?

F 432

G 1,728

H 5,184

J 10,368

15 Use the models to subtract the polynomials. What is the difference of $3x^2 - x + 5$ and $x^2 - 3x + 2$?

 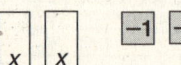 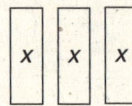

A $4x^2 + 4x + 7$

B $2x^2 + 2x + 3$

C $2x^2 - 2x + 3$

D $3x^2 + 2x - 3$

16 Which expression represents the width of the square below?

$A = 169x^2y^{16}$ } ?

F $169x^2y^{16}$

G $169xy^4$

H $13xy^8$

J $13xy^4$

17 Charlie drew graph #1. He then transformed it to get graph #2. Which transformation did he perform on graph #1 to produce graph #2?

Graph 1

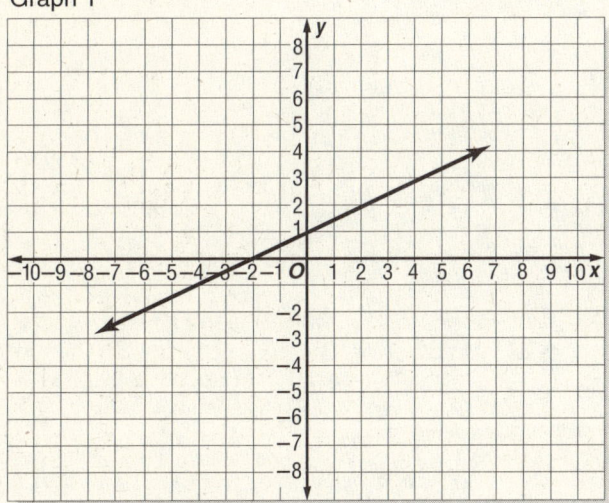

Graph 2

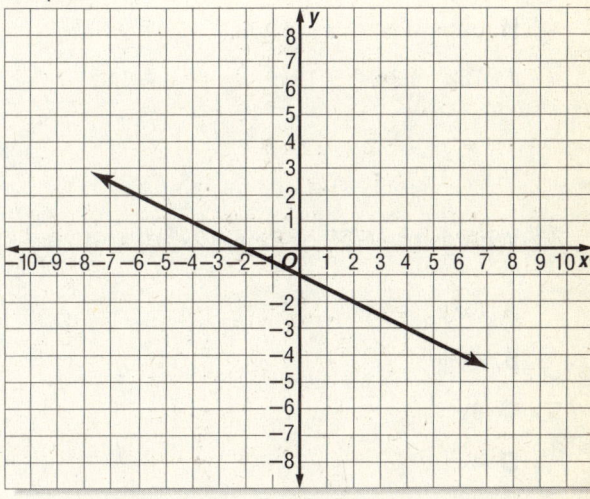

A dilation by a factor of 3

B rotation 90 degrees clockwise

C reflection across the *x*-axis

D translation 2 units downwards

18 A figure is transformed into a new figure by being rotated 180°, translated left 6 units, and dilated by a factor of 4. How are the old and new figures related to each other?

F both congruent and similar

G congruent but not similar

H similar but not congruent

J neither similar nor congruent

19 Which expression does this model represent?

A $3x^2 + 3x + 3$

B $x^2 + x + 1$

C $x^2 - 2x + 3$

D $3x^2 + 2x + 1$

20 What is the quotient of $(6x^5 - 12x^3 - 6x)$ and $3x$?

F $6x^2 - 12x^5 - 6x^2$

G $2x^4 - 4x^2 - 2$

H $6x^4 - 12x^2 - 6$

J $18x^6 - 36x^4 - 18x^2$

21 What is the GCF of $8x^2y$, $12x^4y^4$, and $20x^3$?

A $8x^2y$

B $4x^2y$

C $8x^2$

D $4x^2$

22 Factor $x^2 + 4x - 12$.

    F $(x + 3)(x - 4)$

    G $(x - 3)(x + 4)$

    H $(x + 6)(x - 2)$

    J $(x - 6)(x + 34)$

23 If the following figure is rotated 90° clockwise, what will it become?

A

C

B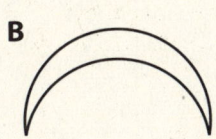

D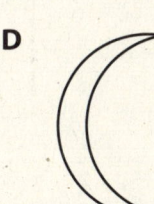

24 Factor $(15a^2bc^2 - 9a^2b^4c)$ by finding the GCF.

    F $3(5a^2bc^2 - 3a^2b^4c)$

    G $3abc(5ac - 3ab^3)$

    H $9a^2bc(2c - b^3)$

    J $3a^2bc(5c - 3b^3)$

25 The chart below describes the height in feet of a boulder hurled from a catapult. For how many seconds does the boulder travel upwards?

t (seconds)	y (feet)
0	0
1	80
2	128
3	144
4	128

A 1 second

B 2 seconds

C 3 seconds

D 4 seconds

26 The width and area of the rectangle are shown in the figure. What is the length?

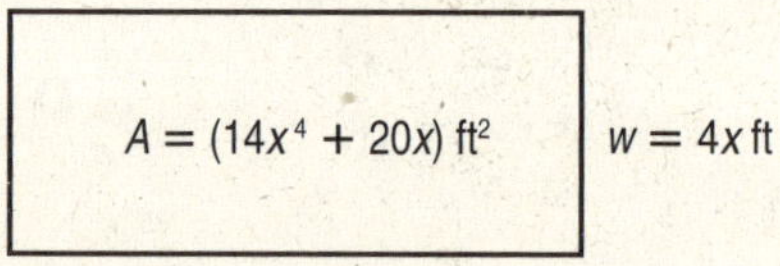

$$A = (14x^4 + 20x) \text{ ft}^2 \qquad w = 4x \text{ ft}$$

F $(3x^3 - 5)$ feet

G $\left(\frac{7x^4}{2} - 5x\right)$ feet

H $\left(\frac{7}{2}x^3 + 5\right)$ feet

J $(3x^4 + 5x)$ feet

27 What is the value of the expression $2^6 \times 4^{-2}$?

A 2

B 4

C 8

D 16

28 A transversal intersects two parallel lines and forms eight angles. Which of the following statements is false?

    F Alternate interior angles are always congruent.

    G Corresponding angles are always congruent.

    H Adjacent interior angles are always supplementary.

    J Exterior angles are always congruent.

29 What relationship do vertical angles have with each other?

    A They are supplementary.

    B They are complementary.

    C They are congruent.

    D They have no relationship.

30 Using the figure, find x.

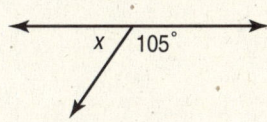

    F 255°

    G 105°

    H 75°

    J $4x^2$

31 A topographic map shows the elevation of a place in feet above sea level. The shaded area shows the parts of a mountain range that are at least 12,000 feet above sea level. If each square contains approximately 10,000 ft² of snow, about how much snow is contained in the shaded area?

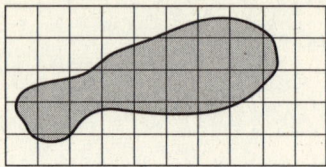

A 10,000 ft²

B 100,000 ft²

C 160,000 ft²

D 1,000,000 ft²

32 Angles *d* and *e* are corresponding angles formed by a transversal and two parallel lines. If $m\angle d = 34°$, what is $m\angle e$?

F 34°

G 56°

H 134°

J 146°

33 Point *A*(2, 5) is reflected over the *x*-axis to point *A′*. What are the coordinates of point *A′*?

A (2, 5)

B (−2, 5)

C (2, −5)

D (−2, −5)

34 Kelly used two intersecting cuts to cut a cake into four pieces, as shown in the diagram below. What is $m\angle x$?

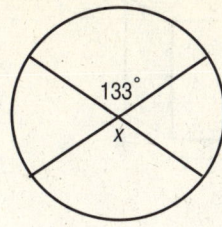

F 43 degrees

G 47 degrees

H 92 degrees

J 133 degrees

35 Angle *a* measures 82°, angle *b* measures 42°, angle *c* measures 98°, and angle *d* measures 8°. Which two angles are complementary to each other?

A angle *a* and angle *b*

B angle *a* and angle *c*

C angle *b* and angle *d*

D angle *a* and angle *d*

36 In the figure below, lines *g* and *h* are parallel. What is the measure of angle *a*?

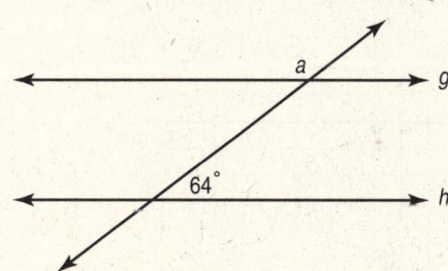

F 64°

G 90°

H 116°

J 124°

37 What is the volume of this box in cubic inches?

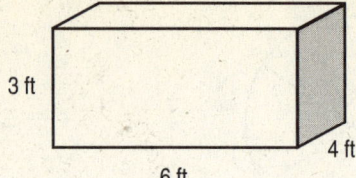

3 ft

6 ft

4 ft

A 72 cubic inches

B 864 cubic inches

C 10,368 cubic inches

D 124,416 cubic inches

38 Yesterday, Judy did p push-ups. Today, she did ten more than half the number of push-ups she did yesterday. Which expression describes the number of push-ups Judy did today?

F $10p + \frac{1}{2}$

G $10p - \frac{1}{2}$

H $\frac{1}{2}p + 10$

J $\frac{1}{2}p - 10$

39 The following figure is reflected over the line $y = x$. What will be the coordinates of point A' of the reflected image?

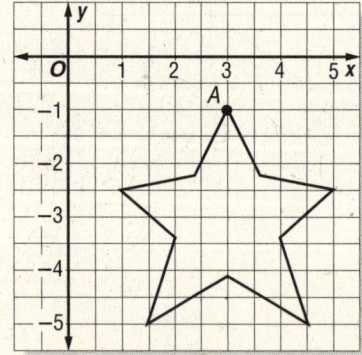

A $(1, 3)$

B $(-1, 3)$

C $(3, 1)$

D $(-3, 3)$

40 Factor the trinomial $x^2 - 5x + 6$.

F $(x + 3)(x - 2)$

G $(x - 3)(x - 2)$

H $(x + 1)(x + 6)$

J $(x - 1)(x - 6)$

41 Line XY is translated 3 units to the left. If a point on line XY is located at (2, 11), where would the equivalent point be on line $X'Y'$?

A (2, 14)

B (2, 8)

C (−1, 11)

D (5, 11)

42 Which equation describes the line below?

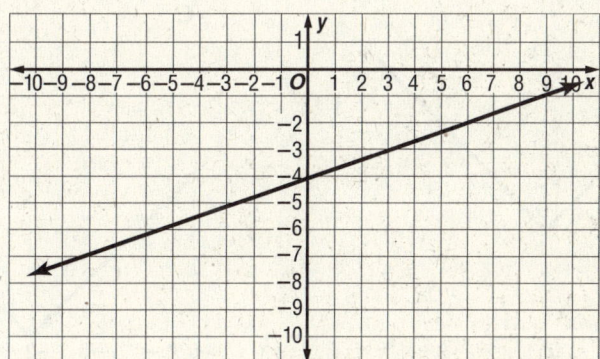

F $y = 3x + 4$

G $y = 3x - 4$

H $y = \frac{1}{3}x + 4$

J $y = \frac{1}{3}x - 4$

43 How many fluid ounces are there in 2 gallons of milk?

    A 256 fluid ounces

    B 128 fluid ounces

    C 64 fluid ounces

    D 8 fluid ounces

44 Anya used the following notation to guide her drawing of a dilation: $(x, y) \rightarrow (5x, 5y)$. What is the size of the dilated figure compared to the size of the original figure?

    F The lengths of the sides are 5 units smaller.

    G The sides are $\frac{1}{5}$ as long.

    H The lengths of the sides are 5 units smaller.

    J The sides are 5 times as long.

45 Which graph is of the function $y = |x + 2|$?

A

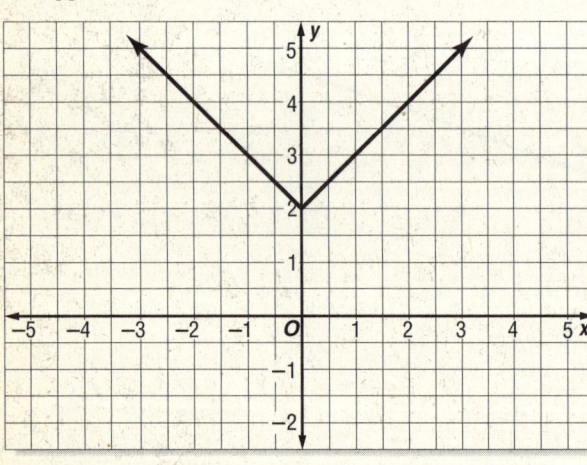

C

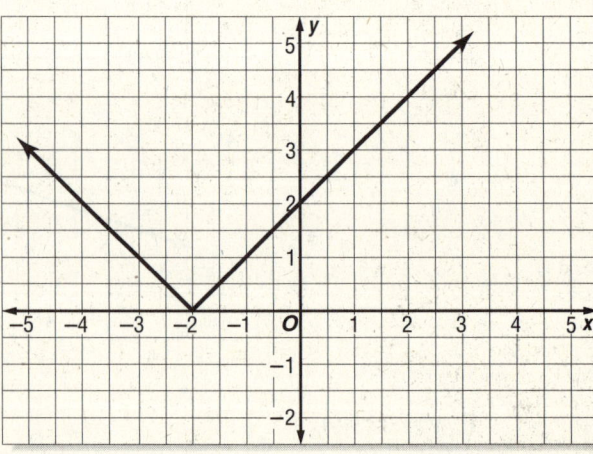

B

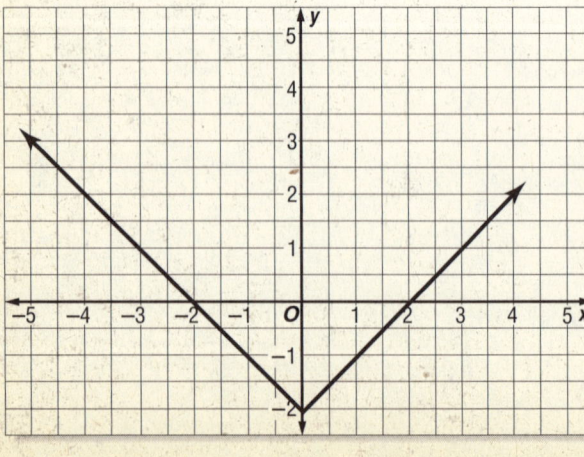

D

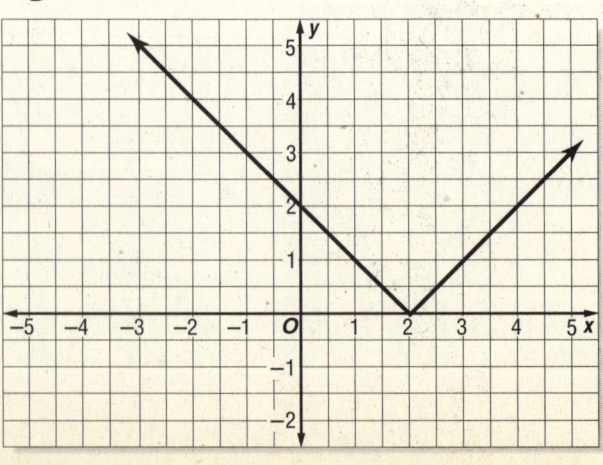

46 A store raised the price of a shirt from $25 to $65. What was the percent increase?

F 60%

G 100%

H 160%

J 260%

47 The area of the figure is shown below. How long is each side?

$A = (x^2 + 6x + 9)$ ft^2

A 3 feet

B $(x + 6)$ feet

C $(x + 3)$ feet

D $(x - 3)$ feet

48 Which transformation has been applied to the solid line to produce the dotted line?

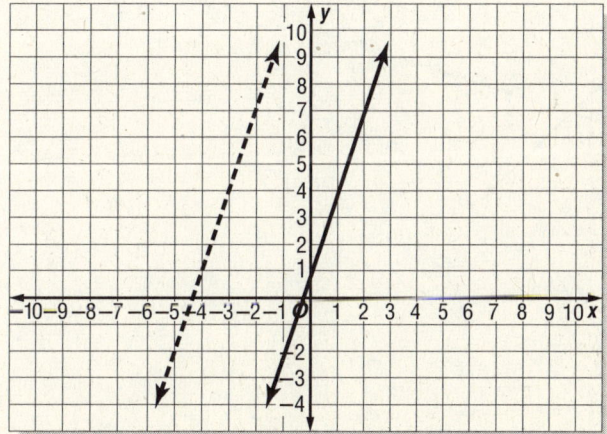

F $f(x + 4)$

G $f(x - 4)$

H $f(x) + 4$

J $f(x) - 4$

49 From the figure below, which equation can you use to find the value of x?

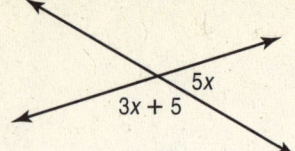

    A $3x + 5 = 5x$

    B $(3x + 5) + 5x = 90$

    C $(3x + 5) - 5x = 180$

    D $(3x + 5) + 5x = 180$

50 Patrick has a summer job passing out flyers for a store. Each flyer weighs about 0.55 ounce. If he is given a bundle of flyers that weighs 16.5 lbs, how many flyers must he pass out?

    F 105

    G 300

    H 480

    J 560

51 Using the figure, find $m\angle a$.

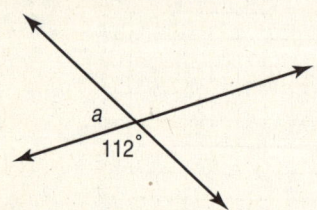

    A $112°$

    B $90°$

    C $72°$

    D $68°$

52 Which set of ordered pairs are on the graph of the function $y = -\frac{1}{2}x - 2$?

    F $(-4, 0), (6, -5)$

    G $(-4, 0), (6, 5)$

    H $(4, 0), (6, -5)$

    J $(4, 0), (6, 5)$

53 Which graph represents the inequality $-3 < x \le 4$?

A

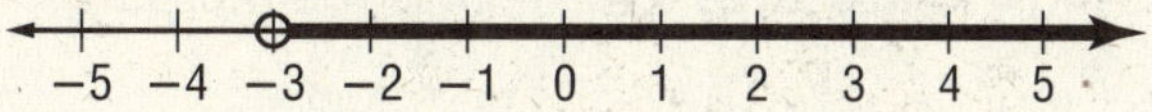

B

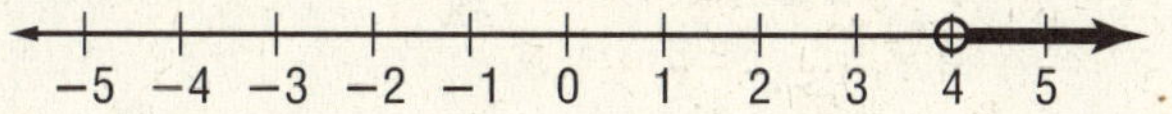

C

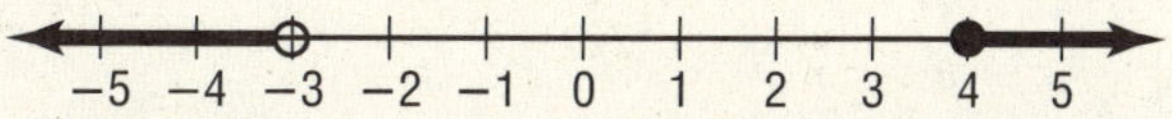

D

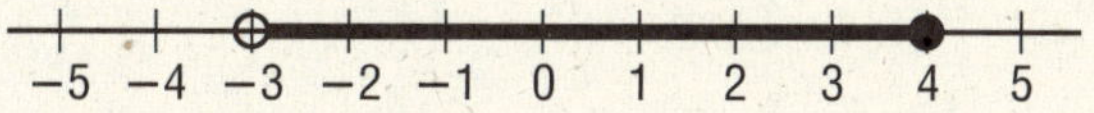

54 Two parallel lines are cut by a transversal. Angles *a* and *b* are adjacent exterior angles. Angle *a* measures 87°. What does angle *b* measure?

    F 3°

    G 87°

    H 90°

    J 93°

55 In a particular year, the amount of salmon in Rock Creek increased by 1,000%. To find 1,000% of a number, you could

A divide by 100

B divide by 10

C multiply by 10

D multiply by 100

56 Angles x and y are complementary. Angle x measures 45°. What is the measure of angle y?

F 145°

G 135°

H 45°

J 15°

57 Which expression gives the area of the rectangle?

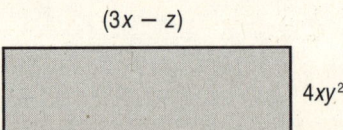

$(3x - z)$

$4xy^2$

A $-12x^2y^2 + 4xy^2z$

B $-1^2x^2y^2$

C $3x + 4xy^2 - z$

D $12x^2y^2 - 4xy^2z$

58 Two parallel lines are cut by a transversal. Angles *a* and *b* are alternate interior angles. Angle *a* measures $(6x + 21)$ degrees, and angle *b* measures $(10x + 9)$ degrees. What is the value of *x*?

F 1

G 3

H 5

J 9

59 A penny is about 1.5 millimeters thick. How many pennies would it take to make a stack 30 meters tall?

A 2,000 pennies

B 20,000 pennies

C 200,000 pennies

D 2,000,000 pennies

60 Simplify $\frac{36xy^3}{8xyz}$.

F $4y^2$

G $18xy^2$

H $\frac{9y^2}{2z}$

J $\frac{9xy}{4z}$

PART 2

61 Francesca kept track of how many ripe apples there were on her apple tree each day. She organized the data in the following chart.

Day	Ripe Apples
1	3
2	5
3	7
4	9

Part A

Plot the data points from the chart, then draw a graph using the points.

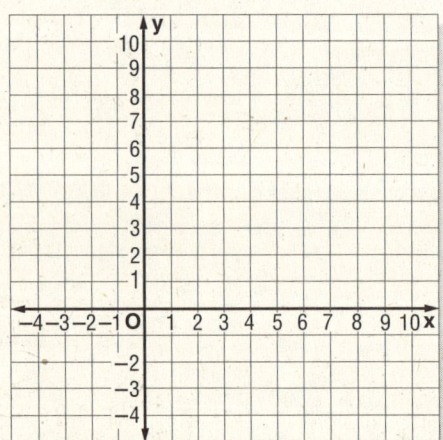

Part B

Write an equation that describes the line you drew in Part *A*.

Answer *y* = _____

Part C

If the same pattern continues, how many ripe apples will there be on Francesca's tree on Day 10?

Answer _____ apples

62 In a parallelogram, consecutive angles are supplementary. In the parallelogram below, $\angle F$ measures $(6x + 1)$ degrees and $\angle G$ measures $(8x + 11)$ degrees. Find the numerical measure of both angles.

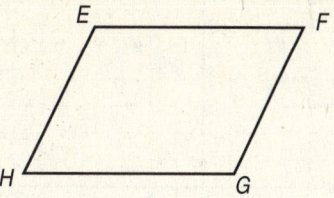

Show your work.

Answer $m\angle F = $ _____ degrees

   $m\angle G = $ _____ degrees

63 Sneakers were on sale for 25% off their original price. Lola bought a pair of sneakers that had an original price of $43.00. She also paid a sales tax of 7.25% on the sale price of the sneakers. How much did Lola pay for the pair of sneakers? Round to the nearest cent.

Show your work.

Answer $ _____

64 Look at the transformation illustrated below.

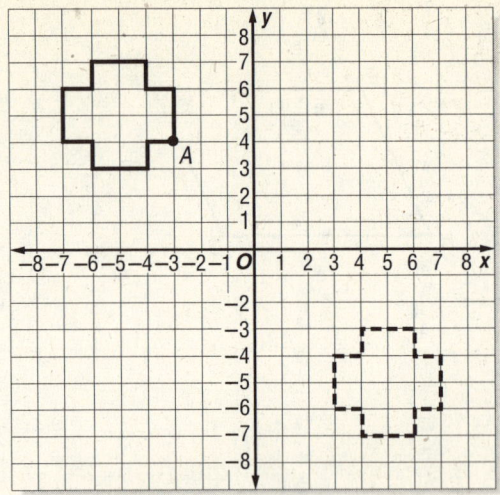

Part A

Two different transformations could have created the figure above. In your own words, describe both of them. Be specific.

1. _____

_____

2. _____

_____

Part B

Look at point *A* on the original figure. Calculate the coordinates of point *A'* using each of the transformations you described in Part A. Write each ordered pair below.

Show your work.

Answer   Transformation 1: Point *A'* = _____

Transformation 2: Point *A'* = _____

65 In the following models, each large square stands for x^2, each bar stands for x, and each small square stands for 1. Shaded figures stand for negative quantities.

Model 1

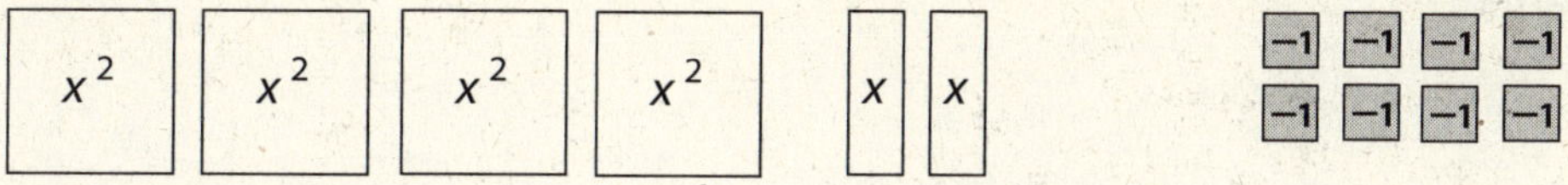

| x^2 | x^2 | x^2 | x^2 | | x | x | | -1 -1 -1 -1 |
| | | | | | | | | -1 -1 -1 -1 |

Model 2

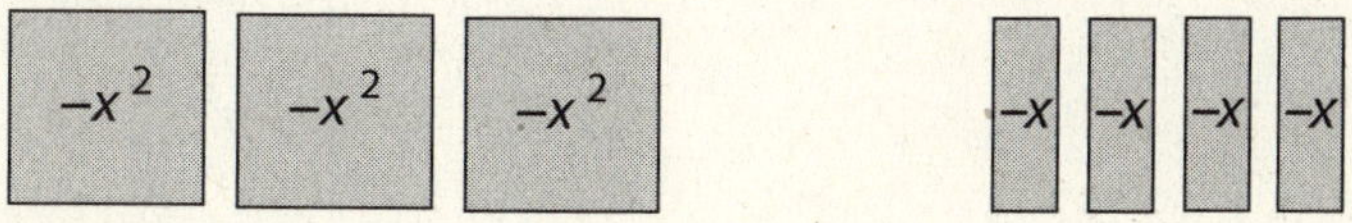

| $-x^2$ | $-x^2$ | $-x^2$ | | $-x$ $-x$ $-x$ $-x$ |

Part A

Translate these two models into mathematical expressions, then add them together.

Show your work.

Answer _____

Part B

Factor the sum you found in Part A. Write the factors and draw a model for each.

Factor 1: _____   Factor 2: _____

Model: Model:

66 Damien sets up a bucket to catch raindrops. The bucket can hold 1 gallon. If each raindrop is made of about $\frac{1}{90}$ fluid ounce of water, how many raindrops will the bucket be able to hold?

Show your work.

Answer _____ raindrops

67 Mr. Periwinkle owns a ranch. He wants to add an area in the shape of an equilateral triangle to one of his rectangular fields, as shown below. If the length of the rectangular field is $3x$ feet and its area is $(9x^2 - 6x)$ square feet, write an expression for the amount of fencing needed to make the new area. Keep in mind that only two sides of the new field need to be fenced in.

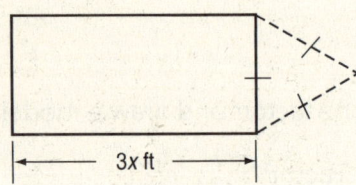

3x ft

Show your work.

Answer _____ feet

68 In the following figure, $\ell \parallel m$.

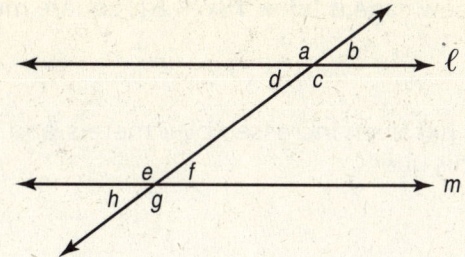

Angle *a* measures $(8x - 10)$ degrees, and angle *f* measures $(3x - 8)$ degrees. Find the numerical degree measure of every angle in the figure.

Show your work.

Answer $m\angle a = $ _____ $m\angle e = $ _____

$m\angle b = $ _____ $m\angle f = $ _____

$m\angle c = $ _____ $m\angle g = $ _____

$m\angle d = $ _____ $m\angle h = $ _____

69 The old lion cage at the zoo was a square, with sides x meters long. For part of the zoo's 50th-birthday celebration, the zoo expanded the length and width of the cage. The expression for the total area of the new cage is $(x^2 + 18x + 81)$ square meters.

Part A

Robert thinks that the cage's length has been increased by 6 meters, and its width has been increased by 8 meters. Check his guess.

Show your work.

Is Robert right? Explain why or why not.

Part B

By how much did the zoo expand the length and width of the cage?

Length: _____ meters   Width: _____ meters

Part C

If the area of the old lion cage was 25 square meters, what is the area of the new cage?

Answer _____ square meters

70 Look at the following diagram of two triangles made by intersecting line segments.

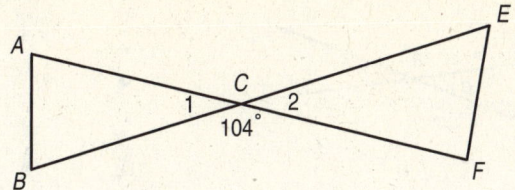

PART A

Find the measures of angles 1 and 2.

Show your work.

Answer $m\angle 1 = $ _____ degrees

    $m\angle 2 = $ _____ degrees

PART B

Triangle *ABC* is an isosceles triangle, with $\angle A \cong \angle B$. Triangle *CEF* is a right triangle, with $m\angle F = 90°$. Using this information, find the measures of the remaining angles in the two triangles.

Show your work.

Answer $m\angle A:$ _____ degrees

    $m\angle B:$ _____ degrees

    $m\angle E:$ _____ degrees

71 The measure of angle *a* is (3*x* + 10) degrees. The measure of angle *d* is (*x* + 10) degrees. What is the value of *x*?

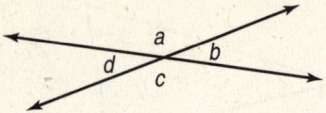

Show your work.

Answer   *x* = _____

72 Glen is making a long chain out of paper clips. Yesterday, he added *p* paper clips to the chain. Today, he added (12*p* −30) paper clips.

Write a verbal expression for how many paper clips Glen added to the chain today.

73 Harold is looking at a map of his backyard. A pond is marked on the map.

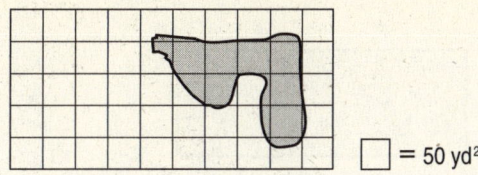

$\square$ = 50 yd²

Part A

Using the diagram above, estimate the area of the pond.

Estimate _____ square yards

Part B

Using your estimate, calculate what percent of the total area of Harold's backyard is covered by the pond.

Show your work.

Answer _____ percent

74 The drawing below shows the length and width of a swimming pool. What expression gives the area of the pool? Make sure you simplify the expression.

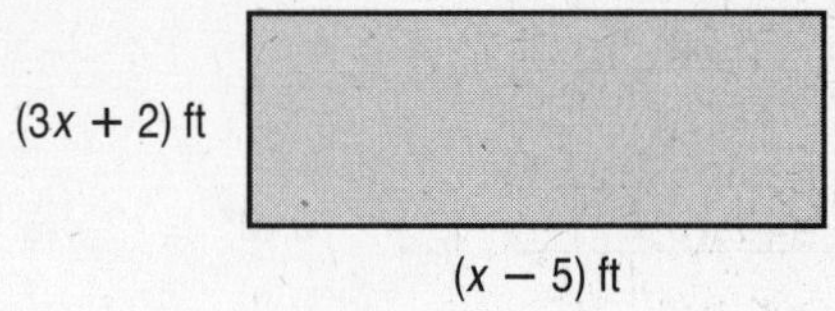

$(3x + 2)$ ft

$(x - 5)$ ft

Show your work.

Answer _____ square feet

75 A line segment is dilated by a factor of 5, then it is translated to the left by 6 units. On the original line segment, endpoint X was at $(4, 2)$ and endpoint Y was at $(-2, -4)$. Where are the equivalent endpoints on the new line segment located?

Show your work.

Answer X'' _____

   Y'' _____

76 Evaluate the following expressions.

Show your work.

$3^3 \times 5^2 + 6^2$

$3^3 \times (5^2 + 6^2)$

Answer _____

_____

Do the expressions have the same value? On the lines below, explain why or why not.

77 The Pleasant Valley Library is moving to a larger building. To get ready for the move, the staff is packing all of the books into boxes. The library owns 35,000 books. On average, each book measures 8 inches by 10 inches by 2 inches. Each box measures 2 feet by 2.5 feet by 1.5 feet.

Part A

About how many boxes will it take to store all of the books? Round to the nearest box.

Show your work.

Answer _____ boxes

Part B

The library uses a truck to carry the boxes of books to the new location. The cargo space in the truck measures 1.5 yards by 4 yards by 2.5 yards. How many trips will it take for the truck to carry all the books to the new building?

Show your work.

Answer _____ trips

78 On the grid below, draw and label the reflection of △*ABC* over the *x*-axis.

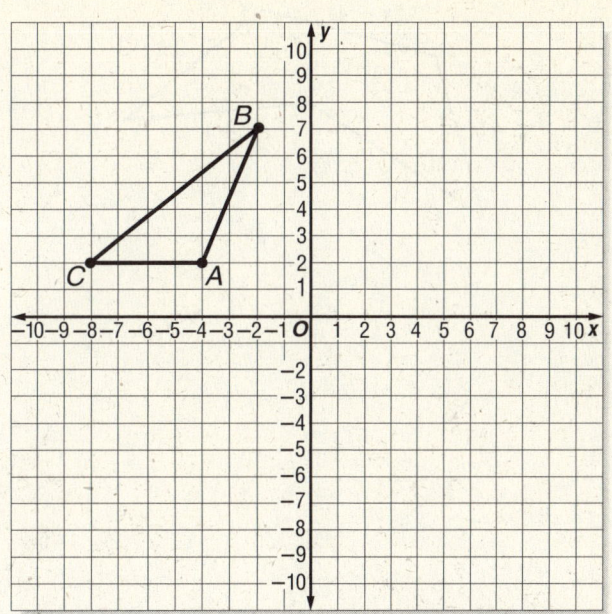

List the coordinates for the vertices of △*A'B'C'*.

Answer   *A'* _____

              B' _____

              C' _____

79 The perimeter of the polygon below is $9x - 1$. Find the expression for the missing side length.

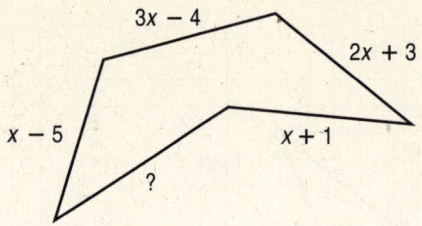

Show your work.

Answer _____

80 Simplify the expression $6x \times 2x^3y \times \dfrac{3x^2y^2}{3x^3} \times x^4y.$

Show your work.

Answer _____

81 Consider the following binomials.

$$A = (x^2 + 5x) \qquad\qquad B = (6x + 30)$$

Part A

Factor each binomial by finding its GCF. Then, add the two factored binomials together to make a single expression.

Show your work.

Answer Factored A + Factored B = _____

Part B

Now add the original forms of binomial A and binomial B together to make a trinomial. Factor the trinomial.

Show your work.

Answer _____

Part C

Do you see something you could do to your answer in Part A to get your answer in Part B? Explain.

_____

_____

_____

_____

82 A hot air balloon goes on a short trip into the air. The relationship between the balloon's height and the time is described by the equation

$$y = -x^2 + 6x,$$

where x is the time in seconds since the balloon took off, and y is the balloon's height in meters.

Part A

Using the equation above, fill in the chart to show the balloon's height.

Time in seconds (x)	Height in meters (y)
0	
1	
2	
3	
4	
5	
6	

Part B

Plot the points from the chart in Part A on the graph below. Then, draw a curve that connects the points you've plotted.

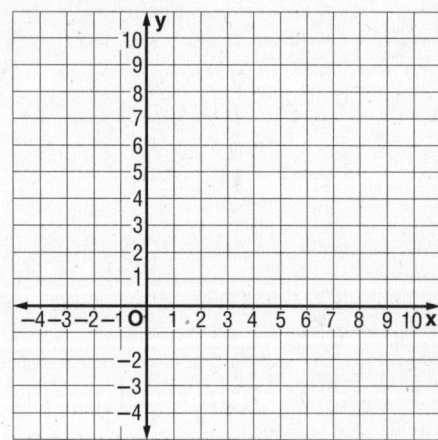

83 Jermaine goes to the store to buy some school supplies. He buys 3 notebooks at $1.49 each, a box of pens for $3.99, and four pencils for $0.29 each. He pays for his supplies with a $20 bill and gets $7.38 back in change. Use estimation to check whether Jermaine got a reasonable amount of change or not.

Show your work.

Did Jermaine get the right amount of change? Explain how you know.

_____

_____

_____

_____

84 Marcus draws two right triangles on a piece of graph paper, Triangle *X* and Triangle *Y*. Both are isosceles right triangles. Triangle *X* is located in Quadrant I, and Triangle *Y* is located in Quadrant III. Each leg of Triangle *X* measures 4 units, and each leg of Triangle *Y* measures 6 units. The hypotenuse of Triangle *X* is facing downwards, while the hypotenuse of Triangle *Y* is facing to the right.

Are the two triangles congruent to each other? Are they similar?
Explain your answer.

_____

_____

_____

_____

_____

85 Toph has a $10 bill and a dime in his pocket. He wants to buy lemon drops at the candy store. Lemon drops cost $0.04 each. He needs to have at least $1.50 left over to take the bus home.

Part A

Write and solve an inequality to find the number of lemon drops Toph is able to buy.

Show your work.

Answer _____

Part B

Graph your answer on the number line below.

$$\xleftarrow{\hspace{1cm}} \underset{0}{|} \quad \underset{25}{|} \quad \underset{50}{|} \quad \underset{75}{|} \quad \underset{100}{|} \quad \underset{125}{|} \quad \underset{150}{|} \quad \underset{175}{|} \quad \underset{200}{|} \quad \underset{225}{|} \quad \underset{250}{|} \xrightarrow{\hspace{1cm}}$$

86 Priscilla performs a series of transformations on a figure in the plane. First, she dilates it by a factor of 3 and translates it 5 units to the right. Next, she reflects it over the x-axis, then over the y-axis. Finally, she reflects it over the line $y = x$, translates it 5 units to the left, and dilates it by a factor of $\frac{1}{3}$.

The point A (4, 3) is on Priscilla's figure. Perform each of Priscilla's transformations described above on point A to find the ordered pair for point A'.

Show your work.

dilation: _____

translation: _____

reflection over x-axis: _____

reflection over y-axis: _____

reflection over $y = x$: _____

translation: _____

dilation: _____

Answer Point A' _____

Glossary

A

adjacent angles (193) Two angles with the same vertex and a common side between them.

algebraic expression (1) An expression that contains operations with variables and numbers.

algebraic pattern (18) A pattern in which the same operation(s) is performed on each number to get the next number in the sequence.

alternate exterior angles (205) When parallel lines are cut by a transversal, a pair of angles formed on opposite sides of the transversal and outside the parallel lines.

alternate interior angles (205) When parallel lines are cut by a transversal, a pair of nonadjacent interior angles formed on opposite sides of the transversal.

B

base (95) A number or variable that is multiplied by itself, or raised to a power given by the exponent; (112) In a percent proportion, the original or whole amount that the part is being compared to.

binomial (163) A polynomial with two terms.

C

clockwise (224) Describes a turn that follows the same direction as the hands of a clock.

complementary angles (198) Two angles whose measures add up to 90°.

congruent angles (193) Angles that have the same measure.

conjecture (110) An educated guess that is based on patterns of results.

constants (1) Terms not multiplied by variables, such as 3 and 11.

coordinate plane (141) Formed when two number lines (called axes) intersect at right angles at their zero points; also called the coordinate system.

corresponding angles (51) The matching angles of similar figures; (204) When parallel lines are cut by a transversal, a pair of nonadjacent angles formed on the same side of the transversal.

corresponding sides (51) The matching sides of similar figures.

counterclockwise (224) Describes a turn that follows the opposite direction of the hands of a clock.

currency (66) Paper money; in the U.S., the unit of currency is the dollar ($).

customary measurement (255) The measurement system used in the United States.

D

diagonal (23) A line segment that joins two nonconsecutive vertices in a polygon.

dilation (218) Enlarging or reducing the size of a figure without changing its form or shape.

E

equation (6) A mathematical statement with an equals sign.

estimation (117) Finding an approximate amount instead of an exact amount.

exchange rate (66) A rate that compares one unit of the currency of one country to the currency of another country.

exponent (95) The number in a power that shows how many times the base is used as a factor.

exterior angles (204) Angles formed on the outer side of parallel lines that are cut by a transversal.

F

FOIL (163) A method for multiplying two binomials. Using FOIL, the First terms in each binomial are multiplied, then the Outer, Inner, and Last terms.

full turn (224) A 360° rotation about a point.

function (28) A relationship in which each input has exactly one output; (146) A special relation in which each element x of the domain set is paired with exactly one element y of the range set.

function rule (28) The operations applied to the input, or the equation that represents a function.

G

greatest common factor (GCF) (176) The largest factor of a set of numbers, or the largest number that divides into each number in a set without a remainder (also called the greatest common divisor).

H

half turn (224) A 180° clockwise or counterclockwise rotation of a figure around a point.

hypotenuse (37) The side of a right triangle that is opposite the right angle and is the longest side in the triangle.

I

image (218) The figure that results from a transformation.

inequality (131) A mathematical sentence that uses the symbols $>$, $\geq$ (greater than or equal to), $<$, or $\leq$ (less than or equal to) to compare two amounts or expressions.

integral exponent (100) An exponent that represents repeated multiplication.

interior angles (23) The angles inside a polygon; (204) Angles formed inside parallel lines that are cut by a transversal.

intersecting lines (193) Lines that cross each other at one and only one point.

L

laws of exponents (95) The rules for multiplying or dividing powers with the same base and integral exponents. To multiply two powers with the same base, add the exponents. To divide two powers with the same base, subtract the exponents.

legs (37) The sides of a right triangle that are adjacent to the right angle.

like terms (1) Monomial expressions that have the same variables, such as $5y$ and $4y$ or $8x^3$ and x^3.

line of symmetry (218) The line that the figure is flipped over in a reflection.

linear pair (193) Two adjacent angles whose noncommon sides are opposite rays.

M

metric units (255) The units of measurement used in most countries (except the United States) and by scientists all over the world. They include grams, meters, and liters.

monomial (1) An expression that is a number, a variable, or the product of numbers and variables.

N

numerical coefficient (158) A number that is multiplied by a variable, usually written in front of the variables: 3 is the coefficient in the term $3x$.

O

order of operations (100) The order in which operations should be performed when evaluating expressions. Working from left to right, first evaluate operations in parentheses, then simplify powers, multiply and divide, and add and subtract.

P

parabola (186) The resulting curve from a graph of a quadratic equation.

parallel lines (204) Lines in the same plane that never intersect.

part (112) In a percent proportion, the amount being compared to the whole amount.

percent (105) The ratio of a part of a whole divided into hundredths to 100; per hundred.

percent decrease (105) The ratio of the amount of decrease to the original amount, written as a percent.

percent increase (105) The ratio of the amount of increase to the original amount, written as a percent.

polygon (23) A closed plane figure that is formed by three or more sides.

polynomial (1) An algebraic expression that contains one or more monomials (called terms).

power (95) A number that can be expressed using an exponent.

principal (113) The base amount of money used to calculate simple interest.

properties of a figure (246) A figure's shape and size, as related to its side lengths and angle measures.

proportion (51) An equation that sets two ratios equal.

Pythagorean Theorem (42) In a right triangle, the square of the length of the hypotenuse (c) is equal to the sum of the squares of the lengths of the legs (a and b), or $c^2 = a^2 + b^2$.

Q

quadratic equation (186) A trinomial equation of the form $y = ax^2 + bx + c$, where a is not zero.

R

rate (56) A ratio that compares amounts in different units, such as miles to hours.

reflection (218) Flipping a figure or finding a mirror image of a figure over a line or over a point.

relation (146) A set of ordered pairs (x, y).

right triangle (37) A triangle that contains a 90°-angle.

rotation (218) Turning a figure around a fixed point.

rule (11) Explains how to find each term in a sequence.

S

scale factor (51) The ratio of a length on a scale drawing or model to the corresponding, real length; (240) For the dilation of a polygon, the ratio of the length of one side of the dilated image to the corresponding side of the original figure.

sequence (11) An ordered list of numbers that follow a pattern.

similar figures (51) Figures that have the same shape but not necessarily the same size.

simple interest (113) The amount of money that is paid or earned for using money. Simple interest is calculated using the formula $I = prt$, where I is the interest amount, p is the principal or original amount, r is the interest rate, and t is the time in years.

solution (6) A value that makes the equation true.

solution set of an inequality (131) The range or set of ranges of values that make an inequality true.

supplementary angles (198) Two angles whose measures add up to 180°.

T

term (1) A monomial in an algebraic expression; (11) Each number in a sequence.

transformation (218) A change in the location, size, or orientation of a figure.

translation (218) Sliding a figure to a different location in the plane.

transversal (204) A line that intersects two or more lines at different points.

trinomial (163) A polynomial with three terms.

U

unit price (56) A unit rate that gives the cost of one item or measured amount, like dollars per pound ($/lb).

unit rate (56) A ratio that compares an amount to 1 unit of another amount, such as distance covered in 1 hour.

V

variable (1) A symbol or letter that represents an unknown quantity in an expression; (141) In a graph, a quantity or amount that changes.

vertex (23) In a polygon, a point where two sides intersect or meet.

vertical angles (193) The opposite angles that are formed by intersecting lines and are equal in measure, or congruent.

Z

zero pair (153) In a polynomial model, two like terms of opposite sign, represented by a positive tile and a negative tile.

Index

Protractor and Ruler Cut-Outs

Cut out and use these tools to help you solve problems that require a ruler 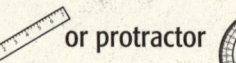 or protractor  .

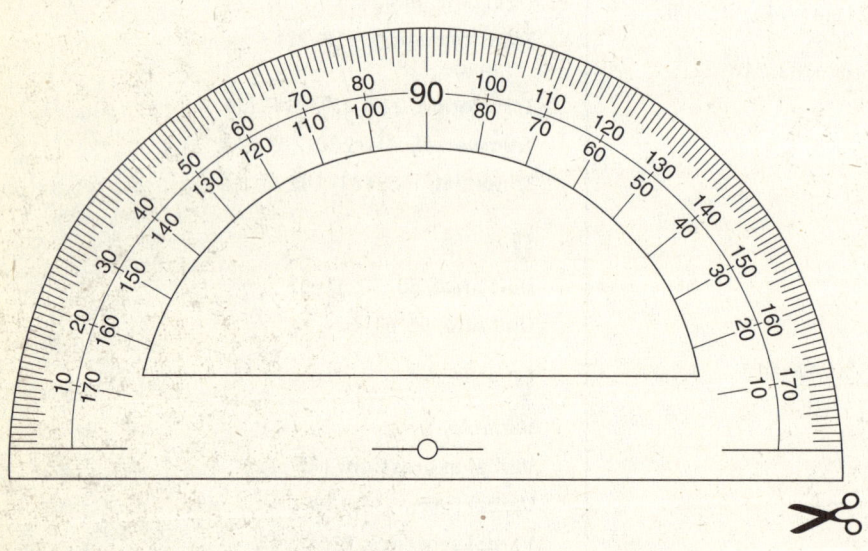